Residential Contracting

Hands-on Project Management for the Builder

August W. Domel, Jr.
Luigina Petrucci

McGraw-Hill, Inc.

New York San Francisco Washington, D.C. Auckland Bogotá
Caracas Lisbon London Madrid Mexico City Milan
Montreal New Delhi San Juan Singapore
Sydney Tokyo Toronto

pbk   1 2 3 4 5 6 7 8 9 0 MAL/MAL 9 9 8 7 6 5 4

Library of Congress Cataloging-in-Publication Data
Domel, August W. (August William), 1960-
    Residential contracting : hands-on project management for the
builder / by August W. Domel, Luigina Petrucci.
        p.   cm.
    Includes index.
    ISBN 0-07-911916-6 (pbk.)
    1. House construction.   2. Contractors.   I. Petrucci, Luigina.
    II. Title.
    TH4812.D66   1994
    692'.8—dc20   94-19862
                                                       CIP

Acquisitions editor: April Nolan
Editorial team: Joanne Slike, Executive Editor
                David M. McCandless, Managing Editor
                Jim Gallant, Book Editor
Production team: Katherine G. Brown, Director
Design team: Lisa Lehocky, Designer   9119116

Table of Contents

Acknowledgments

The authors dedicate this book to their parents Gus Domel, Sr., Jean Domel, Antonio Petrucci and Ines Petrucci. Their sacrifice of personal comfort to finance our education as well as their encouragement in our personal endeavors is greatly appreciated.

The authors would like to thank Roman Szczesniak of Woodridge, Illinois for his help in reviewing the drawings. Also, we would like to acknowledge the helpful input provided by Tim Kral and Janet Kral of T. T. Kral Builders in St. Charles, Illinois. Their commitment to high quality construction before profits is truly an inspiration to those in the industry.

Last, but not least, we would like to thank Lisa Lehocky, of Desktop Creations, in Round Lake Park, Illinois. Lisa single-handedly created every drawing as well as designed and prepared this entire text. More importantly, she made revision, after revision, after revision of the drawings without a complaint. We sincerely thank her.

Introduction

A great sense of accomplishment is felt upon the completion of a construction project. This feeling is present for projects as small as the assembly of an airplane model or for projects as large as the construction of a high-rise. One takes great pride in knowing that a pile of material can be assembled to form a complete unit. These feelings are enhanced when the construction project is the building of a home. Satisfaction of a technical accomplishment is enhanced by knowing the finished product will provide shelter and safety for a family. Residential contractors rate their success not only on profit, but also on customer satisfaction.

A general contractor must function not only as a construction expert, but also as an accountant, mediator, material specifier, coordinator, marketer, to name a few. This publication teaches the fundamentals needed to be a successful residential contractor. However, the reader is warned that reading a book on construction is like reading a book on bicycle riding: one never really learns until one actually tries it. This book tries to reduce the number of falls.

This book is written for those interested in entering the residential contracting field. Prior experience in the construction field is helpful, but by no means necessary. This book is also written for those who are already residential contractors, but would like to fine tune their knowledge.

The most important aspects of residential contracting are presented in the following five chapters. A brief summary of what the reader can expect is given in the remainder of this chapter.

Chapter 1—Land Zoning and Construction Permits

This chapter introduces the reader to the world of land zoning. Terminology and common problems are presented. Sample construction permits are also provided.

Chapter 2—Home Building Construction Scheduling

A detailed construction schedule is provided. Each item on the schedule is discussed in detail. Included in this discussion is completion time, percent of total cost, scope of work, and other useful information.

Chapter 3—Construction Contracts

This chapter dissects a complete construction contract. Each clause of the contract is explained. A check list is provided at the end of the chapter to assist in contract review.

Chapter 4—Building Plans

A unique step-by-step procedure is used to completely explain a set of residential building plans. Photographs and sketches are included to clarify details.

Chapter 5—Specifications

This chapter presents and explains the most common items included in construction specifications. Forms to aid in the organization of material selection are provided.

The amount of material given for the reader to digest may seem overwhelming. It is important to realize that although general contracting is a rewarding career, it is also detail oriented. The reader should take the time to review this publication more than once to gain its full value.

Chapter 1

Land Zoning and Construction Permits

Many projects die a slow death while the general contractor tries to obtain a land zoning change or tries to get the necessary permits. Poor planning by the general contractor, and delays by overworked building departments, result in more time spent seeking a permit than it takes to construct the house. Requests for land zoning changes can be equally frustrating as time slips away and legal costs accumulate.

This chapter provides the reader with a working knowledge of land zoning and permits. The general contractor must be knowledgeable in land zoning when building an entire subdivision, when the project is conditioned on a zoning change, or when the contractor is purchasing a parcel of land and selling the house and land as one package.

The various land zoning issues presented in this chapter are:

- The Zoning Ordinance
- Spot Zoning
- Conditional Zoning
- Variances
- Special Use
- Non-conforming Use
- Aesthetic Zoning
- Planned Unit Development

Various types of permits are also discussed.

The Zoning Ordinance—The zoning ordinance divides a city into different land zoning districts. Each land zoning district will place similar land parcels in similar areas into the same classification. The number of classifications depends on the variety of businesses, population density, and common land use in the area. A sample of the possible classifications that could be used in a suburban setting follows.

R1	Single family residential district
R2	Single family residential district
R3	Residential transition district
R4	Multi-family residential district
C1	Limited commercial and office district
C2	Shopping center district
C3	Highway commercial district
I	Industrial district
AG	Agriculture district

The zoning ordinance should be consulted for each city since the designations and terminology may have different meanings. The allowable uses for the classifications listed above may be as follows:

R1—Single Family Residential

This district is designed to provide for a low-density living environment. Only single-family detached dwellings are permitted in this district.

R2—Single Family Residential

This district is designed to allow for large lots. Only single-family detached dwellings are permitted in this district.

R3—Residential Transition

This district is designed to provide for a medium-density living environment. Structures are limited to single-family detached, two-family dwellings, and townhouses.

R4—Multi-family Residential

This district is designed for a planned mixture of residential dwellings limited to 10 dwellings per acre.

C1—Limited Commercial and Office

This district is limited to small commercial and office buildings including book stores, florist shops, gift shops, ice cream shops, and similar office uses including architects offices, insurance agencies, newspaper offices, employment agencies, and similar businesses.

C2—Shopping Center

This district is designed for large neighborhood shopping centers. Shopping center use includes carpet stores, department stores, drug stores, toy stores, television sales, shoe stores, and similar businesses.

C3—Highway Commercial

This district is designed to provide a variety of commercial facilities. Highway commercial use shall include automobile sales and services, appliance stores, veterinary clinics, medical clinics, real estate offices, and similar businesses.

I—Industrial

This district is designed for warehouse, office, and research activity. Industrial districts include contractor offices, book keeping services, testing and research offices, laboratories, and restaurants.

AG—Agriculture

This district is designed for farming use.

Spot Zoning—The governing body of a locale provides a zoning scheme that reasonably divides the property into different zoning classifications. Each classification consists of the grouping of very similar adjacent parcels of land. It is not surprising that some individuals will later request that the parcel of land be rezoned for a more advantageous use. Such a zoning change is welcomed so the highest and best use of the land is obtained, provided that the change is in line with the overall zoning scheme. A change that results in zoning that is not in harmony with the overall plan is called spot zoning. If 50 parcels of property are zoned for residential development and one is rezoned for heavy industry, spot zoning may be occurring. A zoning change is considered spot zoning if it is a drastic change not in accordance with the overall plan, and affects a small parcel of land for the benefit of only the parcel owner. Spot zoning will be prohibited by the court since allowing such a change undermines the zoning system, detrimentally affecting adjacent land owners and invites graft.

Conditional Zoning—Spot zoning is an outright, no strings attached, zoning change. In contrast, conditional zoning is a rezoning of a piece of property with a requirement that something be done before the zoning change is allowed. For example, a general contractor may desire to build multi-family homes on land zoned for single family use. The zoning board may grant the change on the condition that the existing streets in the neighborhood be widened to accommodate the increased traffic. Such conditional zoning is frowned upon by the courts since it invites "behind closed door," deals and there is no established criteria for allowing the change.

Variances—Allowing a parcel of land to have different zoning classifications than similar adjacent property is in violation of a fair zoning ordinance. However, to effectively function, the zoning laws must have at least some flexibility. Spot zoning and contract zoning are changes given by the zoning board just for the purpose of a more advantageous classification. A variance on the other hand allows for a more advantageous classification because the owner has a unique problem.

Variances are of two types, "use" and "area." The use variance allows the parcel to be used in a manner inconsistent with the zoning ordinance. The area variance allows for the relaxation of physical dimensioning requirements (minimum house size, minimum distance from roadway, etc.). A variance is only allowed if the three following requirements are satisfied.

- Present zoning results in an unnecessary hardship to the owner.

- The problem is unique to this parcel of land.

- The variance is not totally inconsistent with the overall zoning plan.

Special Use—In some cases, a parcel of land will have a zoning classification plus an additional zoning option. For example, a parcel of land may be zoned for single family use with an option to allow a church, school, or hospital. This additional zoning option is called a special use. A special use zoning request is not a change in zoning, but still needs approval of the zoning board. The need for approval arises because, although hospitals and schools are beneficial to the public, there are serious ramifications on traffic and adjacent parcels. The special use allows the zoning board to have some control over the land development. Note that the special use is different than a variance, since a variance requires that the owner be suffering a hardship.

Non-Conforming Use—When an area is zoned for the first time or even when land is rezoned, some of the existing uses of the land will be in conflict with the new zoning ordinance. Elimination of these existing uses would be unwise and unfair. To avoid this problem, the zoning board allows a non-conforming use. The granting of a non-conforming use allows the owner to continue to use the land for its current use, despite the conflict in zoning. The non-conforming use can be for an indeterminate or limited period of time. The conditions of the zoning relaxation are contingent on the business not being expanded or changed.

Aesthetic Zoning—Some municipalities require that approval from an architectural review board be received prior to construction. Although such reviews seem subjective and can possibly lead to abuse, it has been upheld by the courts.

Planned Unit Development (PUD)—A PUD differs from the typical zoning scheme. A typical zoning ordinance places many land parcels into one zoning classification. A PUD takes the same group of parcels, but allows several zoning classifications. This gives the group of parcels the flexibility to have single family, multi-family, and light commercial interspersed to form a more diverse and convenient neighborhood.

The following permits are required on most non-urban construction projects.

Building Permit Application (see Fig. 1-1)—The building permit provides the official approval to start a project. Multiple copies of the design drawings with a fee are submitted to the building department. The building department reviews the drawings to ensure that they are in accordance with the local building code. If the design drawings are acceptable, a building permit is issued (Fig. 1-2).

Septic Permit (Fig. 1-3)—A permit is required to either lay a septic field or for connection to the sewer system.

Private Well Water Permit (Fig. 1-4)—If the owner is to construct a well on the lot, a permit and a water analysis is required.

Culvert Permit (Fig. 1-5)—Construction of a driveway through a swale will impede the water flow and will necessitate a culvert. Since the size of the culvert has a direct bearing on the water flow, a permit is required.

YOUR COUNTY DEVELOPMENT DEPARTMENT
Building and Zoning Division
APPLICATION FOR CONSTRUCTION

1. Date of application _____

2. Owner or Lessee of property

    Name _____

    Address _____

    City _____   Zip _____

    Telephone _____

3. Mail permit to _____

    _____

    _____

4. Legal description of property (Print, type, or write legal description as it appears on deed).

    _____

    _____

    _____

    _____

5. Indicate below the type and use of building or structure proposed to be constructed.

    _____

    _____

    _____

6. Health Dept. No. _____   Estimated Cost _____

7. Name, Address, Telephone, and License No. of

    General Contractor _____   Phone No. _____

    _____

    Plumbing Contractor _____   Phone No. _____

    _____   License No. _____

Figure 1-1

YOUR COUNTY BUILDING AND ZONING PERMIT

To:
                                           Permit No.

Map No.

Zoning Classification

Fire No.

Your application has been received for a XXX County Building and Zoning permit, dated
in which you state 1) that you are the owner/lessee of the premises legally described as follows:

2) that you desire to make thereon certain described improvements costing   consisting of

and

3) that you agree to use such premises and existing and proposed buildings and structures for only the uses as shown above, under item 2.

The proposed improvements and uses as applied for, are permissible under the terms of the Zoning Ordinance adopted by the Board of Supervisors of XXX County, State, on (specify date), and Building Ordinance adopted (specify date) and subsequently amended and the permission applied for is hereby granted.

Work authorized hereby must be in accordance with the sketch on your application for this permit. Notify the XXX County Building and Zoning Division immediately, of any proposed changes, before such changes are made.

_____   _____
Enforcing Officer of XXX County Zoning Ordinance   Enforcing Officer of XXX County
Building Ordinance

Dated   Dated

Note the following penalties and enforcements provided in state zoning act and county building and zoning ordinance.

Any person, firm, company, or corporation who violates, disobeys, omits, neglects, or refuses to comply with, or who resists the enforcement of any of the provisions of the building and zoning ordinance, shall be subject to a fine of not more than 200 dollars for each offense and to imprisonment in the county jail for a period of not more than six months, or both, at the discretion of the court. Each day that a violation continues to exist constitutes a separate offense. In case of violation in addition to other remedies, any appropriate action or proceedings in equity may be instituted (1) to prevent such unlawful erection, construction, reconstruction, alteration, repair, conversion, maintenance, or use, (2) to restrain, correct, or abate such violation, (3) to prevent the occupancy of said building, or land, or (4) to prevent any illegal act, conduct, business, or use in or about the premises.

Check Zoning and Building Ordinances for time limits in starting and completing work authorized above.

Parcel No.

Figure 1-2

YOUR COUNTY HEALTH DEPARTMENT
Private Sewage Disposal System
APPLICATION

Fee $ _____

CONTRACTOR INFORMATION

Date of Application: _____

Name: _____

Name: _____

Address: _____

Address: _____

Telephone: _____

City: _____ Zip: _____

County License _____

Telephone: _____

State License No. _____

Mail Permit To: _____

Subdivision _____   Unit _____   Lot ____   Block _____

Township _____   Section _____

Legal Description _____

Directions _____

TYPE OF BUILDING

❏ Single-family residence   ❏ Multi-family residence   ❏ Business Industry

PROPOSED CONSTRUCTION
(Check the following information for required permit)

❏ New field & septic tank   ❏ New septic tank only

❏ New field only   ❏ Addition of _____ ft. to field

Residential

Family units _____   Garbage grinder _____

Bathrooms _____   Dishwasher _____

Bedrooms_____   Sinks _____

Number of people _____   Bathtubs _____

Basement _____   Showers_____

It is clearly understood that the owner assumes full responsibility in obtaining the inspection and final approval of the County Health Department on all portions of this sewage disposal system installation prior to covering any portion of this system. In requesting an inspection, call the County Health Department at (555) 555-5555 and give the permit number.

I hereby certify that, to the best of my knowledge, the preceding information is correct. In addition, the sewage disposal system will be installed strictly as outlined in this permit application in conformance with the County Septic Ordinance.

Septic Field Area Has Been Staked and Roped Off   ❏ Yes ❏ No

Sewage System Contractor   Date   Owner   Date

Approved by: _____   Date _____

Figure 1-3

YOUR COUNTY HEALTH DEPARTMENT
Application For Private Well System

1) Property Owner Current Address

        Name: _____

        Address: _____

        City: _____

        Telephone: _____

2) Contractor

        Name: _____

        Address: _____

        City: _____

        Telephone: _____

        License: _____

3) Well Location

        County: _____

        City: _____

        Subdivision: _____

        Lot #: _____

        Legal description: _____

        _____

4) Construction Information

        Well Diameter: _____

        Well Depth: _____

_____   _____

Signature of Subcontractor   Date

Figure 1-4

VILLAGE ROAD COMMITTEE
Driveway Permit

WHEREAS _____ hereinafter termed "PETITIONER", requests permission and authority to put a culvert and driveway in, upon, or along, Highway/Street known as _____ on property located (Subdivision/Legal Description)

subject to the following conditions and restrictions:

FIRST: The PETITIONER shall furnish all material, do all work, pay all cost, and shall restore said highway to a condition similar, or equal to that existing before the commencement of described work. The work permitted here shall be completed before the construction of any structures.

SECOND: The proposed improvement shall be located and constructed to the satisfaction of the Village Road Commissioner, and shall conform to the material and installation standards herein attached. The improvement shall be completed within 30 days of the issuance of this permit, and the Village Commissioner shall be notified upon completion of the improvement in order to inspect the improvement.

THIRD: If during the installation of the culvert or driveway, the roadway is damaged, the PETITIONER shall restore the roadway to the satisfaction of the Village Road Commissioner.

FOURTH: This permit does not release PETITIONER from fulfilling any existing statutes relating to the construction of such improvements.

THIS PERMIT IS HEREBY ACCEPTED AND ITS PROVISIONS AGREED TO THE_____ DAY OF _____ A.D., 19___.

PETITIONER _____   VILLAGE ROAD COMMISSIONER _____

ADDRESS _____

TELEPHONE _____

Figure 1-5

Chapter 2
Home Building Construction Scheduling

The secret to success for a general contractor is proper construction scheduling. Each day that is not used to its fullest potential is a cost to the general contractor. The general contractor must set a goal of providing a high quality product produced in the least amount of time. This goal is reached by using reliable quality subcontractors in conjunction with efficient scheduling practices. The subcontractors must be scheduled in proper sequence so subcontractors can start at the earliest possible date without being in the way of other contractors.

The type of subcontractors needed to build a home are:

- Excavation
- Rough Framing Carpenters
- Heating and Air Conditioning
- Roofing
- Well Drilling
- Siding
- Insulation
- Painting
- Carpet Installers
- Clean Up
- Concrete
- Plumbing
- Electrical
- Masonry
- Septic Installer
- Gutter
- Drywall
- Trim Carpenters
- Tile Installers

The bar chart on the next page shows the percentage cost breakdown for construction of a home. This cost breakdown provides two interesting pieces of information. First, it should be noted that half of the work items account for 90% of the overall cost. The general contractor should spend considerably more time on these high ticket items and should send out for a greater number of bids from these subcontractors. It is more efficient to spend time and effort trying to reduce the high priced item because their reduction has a greater impact on the overall cost. The second piece of important information that is shown in the construction breakdown is that 10 of the work items control two-thirds of the cost. For these items, the following is recommended:

- Request three to four bids from subcontractors.

CONSTRUCTION COST BREAKDOWN

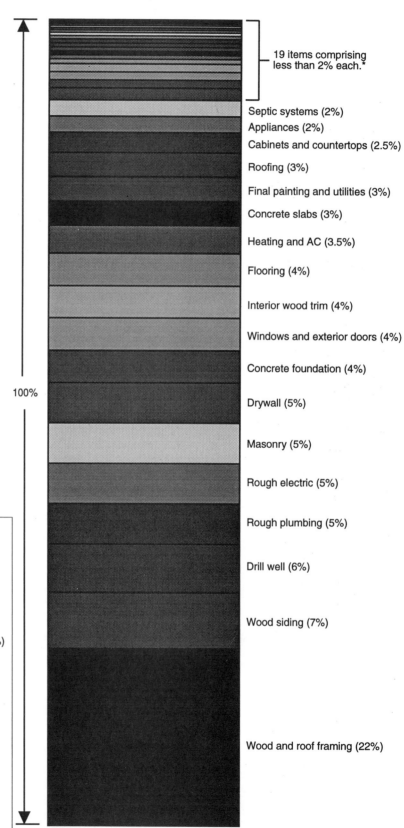

19 items comprising less than 2% each.*

Septic systems (2%)
Appliances (2%)
Cabinets and countertops (2.5%)
Roofing (3%)
Final painting and utilities (3%)
Concrete slabs (3%)
Heating and AC (3.5%)
Flooring (4%)
Interior wood trim (4%)
Windows and exterior doors (4%)
Concrete foundation (4%)
Drywall (5%)
Masonry (5%)
Rough electric (5%)
Rough plumbing (5%)
Drill well (6%)
Wood siding (7%)
Wood and roof framing (22%)

100%

Items under 2%*

Insulation (1.5%)
Finish trim (1%)
Install concrete footings (1%)
Garage doors (1%)
Excavate foundation (0.6%)
Foundation drain (0.5%)
Set steel columns and beams (0.5%)
Prefabricated fireplace (0.5%)
Gutters (0.5%)
Primer (0.5%)
Brace walls and backfill (0.4%)
Clean-up (0.4%)
Mirrors (0.3%)
Culvert (0.3%)
Strip and stockpile soil (0.3%)
Damp proof walls (0.2%)
Stairs and railings (0.2%)
Spread topsoil (0.2%)
Stake-out house (0.1%)

- Research the quality and reliability of the lowest bidder.
- Meet with those subcontractors to investigate cost cutting ideas.

This chapter presents a construction scheduling chart called a Gantt Chart. The construction schedule shows the start and completion date relative to the other tasks. Also shown (denoted by the "I") is the time when an inspection by the building department is required.

The construction schedule provides for a completion time of 12 weeks. The 12 weeks should be a goal, but is not the norm. This time frame does not include the time lost to weather, labor strikes, and other unforeseen acts of God. A four to six month duration for construction is more realistic. However, the successful general contractor should strive to finish by the 12th week.

Each item on the construction schedule is discussed in considerable detail in the remainder of this chapter. The information included on each item is:

- Time to complete task
- Percentage of total cost
- Purpose of the work item
- The work that is needed to complete the item
- Who provides the services and material for each work item
- Alternatives

Another point to consider in keeping a tight schedule is availability of materials. Ordering of material provided by the subcontractor is not of concern to the general contractor. However, there are many instances where by custom the general contractor provides the materials to the subcontractors. Below is a list of materials that are typically provided by the general contractor.

- Steel beams and steel columns
- Framing material
- Window and exterior doors
- Stairs and railings
- Siding
- Cabinets and countertops
- Wood Trim
- Appliances
- Garbage dumpsters

Careful and thorough reading of the remainder of the material in this chapter should provide a good foundation in home building construction scheduling.

GANTT CHART—RESIDENTIAL CONSTRUCTION

Project Name _____

Project No. _____   Date _____

Project Start Date _____

#	Task	WEEK 1	WEEK 2	WEEK 3	WEEK 4	WEEK 5	WEEK 6
1	STAKE OUT HOUSE	▓					
2	INSTALL CULVERT	▓					
3	STRIP & STOCKPILE TOP SOIL	▓					
4	EXCAVATE FOUNDATION	▓					
5	FOOTINGS	▓ (I)					
6	FOUNDATION WALLS & PIER FOOTINGS		▓				
7	INSTALL FOUNDATION DRAIN		▓ (I)				
8	DAMP PROOF WALLS			▓ (I)			
9	BRACE WALLS & BACKFILL			▓ (I)			
10	SET STEEL COLUMN & BEAMS			▓			
11	FRAME—1ST LEVEL—FLOOR			▓			
12	FRAME—1ST LEVEL—WALLS				▓		
13	FRAME—2ND LEVEL—FLOORS				▓		
14	FRAME—2ND LEVEL WALLS					▓	
15	ROOF FRAMING						▓
16	ROUGH PLUMBING						▓
17	HEAT & A.C.						▓
18	TELEPHONE WIRING & ROUGH ELECTRIC						▓
19	WINDOWS & EXTERIOR DOORS						▓
20	ROOFING						▓
21	MASONRY						▓
•	ORDER STEEL		○				
•	ORDER WOOD FRAMING		○				
•	ORDER CABINETS & COUNTERS		○				
•	ORDER APPLIANCES		○				
•	ORDER WINDOWS & DOORS			○			
•	ORDER SIDING			○			
•	ORDER LIGHT FIXTURES					○	
•	ORDER DUMPSTER						○

GANTT CHART—RESIDENTIAL CONSTRUCTION

Project Name _____

Project No. _____

Project Start Date _____

Date _____

	WEEK 7	WEEK 8	WEEK 9	WEEK 10	WEEK 11	WEEK 12
19 WINDOWS & EXTERIOR DOORS (CONT.)						
20 ROOFING (CONT.)						
21 MASONRY (CONT.)						
22 STAIRS						
23 PREFABRICATED FIREPLACE						
24 SLABS						
25 DRILL WELL						
26 SEPTIC SYSTEM						
27 SIDING						
28 GARAGE DOORS						
29 GUTTERS						
30 INSULATION						
31 DRYWALL						
32 PRIMER PAINT						
33 CABINETS/COUNTER TOPS						
34 FINAL PAINT						
35 FINISH ELECTRIC						
36 FINISH PLUMBING						
37 WOOD TRIM						
38 FLOORING						
39 APPLIANCES & A.C. UNIT						
40 MIRRORS						
41 FINISH TRIM						
42 CLEAN UP						
43 SPREAD TOPSOIL						
• ORDER TRIM						
• ORDER MIRRORS & SHOWER DOORS						
• ORDER DUMPSTER						

1—Stake Out House

Completion Time—1 Day
Percent of Total Cost—0.1

Purpose: The boundaries of the house are marked on the property so the excavator can determine where the foundation will be constructed (Fig. 2-1).

Work Included: This task can be performed by the general contractor although it is advised that a licensed surveyor be used. The work includes placing stakes at all locations where corners of the proposed house will be.

Furnishing Material: Minimal material provided by surveyor.

Inspections: None required.

Alternatives: Stake out can be omitted, but problems may occur if structure is improperly located.

Services by: Surveyor.

Figure 2-1 Stake out of property

2—Culvert

Completion Time—1/2 Day
Percent of Total Cost—0.3

Purpose: Water from rain and melted snow that accumulates on drive-ways, roads, and on building property must drain and be collected at the proper location. In an urban area, the water on these surfaces flows into the storm sewer system and to a lake or reservoir. In unincorporated areas, sewer systems are typically not used, since the cost is prohibitive and other methods of runoff collection are more feasible. In unincorporated areas, the runoff water drains from the driveways, roads, and the lot into a swale. A swale is a ditch that is several feet wide and runs along the front of the lot. The swale funnels the water to a designated location, usually a retaining pond.

Since a swale is several feet deep, automobiles cannot traverse the swale to reach the house. Therefore, the swale must have a culvert pipe buried in gravel where the driveway is located. The culvert allows the water to flow virtually unobstructed in the swale while allowing vehicles to travel over the swale and enter the driveway.

Work Included: Provide and install a steel or concrete pipe on a compacted bed of sand or stone at the edge of the driveway but several feet below. The pipe is then buried and covered with rock until it reaches the height of the adjacent roadway.

Furnishing Material: Material provided by excavation contractor.

Inspections: The layout of a water runoff plan is determined by a civil engineer. This plan is usually completed before the tract of land is subdivided into single lots. Prior to construction of the culvert, the pipe must be properly sized and the location approved by the appropriate governing body so it will not alter the planned water flow.

Alternatives: As previously discussed, in urban areas, a storm sewer system, as well as curbs, and gutters are usually in place before the property is subdivided and sold. If this is the case, the culvert is not needed.

Services by: Excavation contractor.

3—Strip and Stockpile Soil

Completion Time—1/2 Day
Percent of Total Cost—0.3

Purpose:	The top few feet of soil is typically black dirt. This soil is fertile and is used for growing grass and other plants. Below this level of black dirt are either clays, sands, and silts, which are not conducive to plant growth. When construction of the house is completed, black dirt is required for the top few feet of the backfill. The cost of purchasing top soil is avoided by stripping the black dirt, stockpiling it, and then spreading it when the construction is complete.
Work Included:	Strip black topsoil from driveway and excavation area and stockpile.
Furnishing Material:	No material needed.
Alternatives:	Do not strip and save topsoil, but purchase new soil when construction is completed.
Inspection:	None required.
Services by:	Excavation contractor.

4—Excavate Foundation

Completion Time—2 days
Percent of Total Cost—0.6

Purpose: The soil in the location where the foundation will be constructed must be excavated.

Work Included: Excavation of complete foundation as per the design drawings and store it in a pile away from construction area, and remove if it is not needed for backfilling. All applicable safety requirements including a proper slope of excavation must be satisfied.

Furnishing Material: No material needed.

Inspections: None required.

Alternatives: Elimination of the basement would considerably reduce the amount of excavating required.

Services by: Excavation contractor.

Figure 2-2 Foundation excavation

5—Footings

Completion Time—2 Days
Percent of Total Cost—1.0

Purpose:	The footings, which spread the load of the wall over a wider area of soil to avoid settlement, must be constructed prior to placing the basement wall formwork.
Work Included:	Soil must be checked to see if it is of suitable quality before any work starts. Work includes placing formwork, pouring concrete, making keyways, and when the concrete has hardened, stripping of the formwork.
Furnishing Material:	Material provided by concrete contractor.
Inspections:	An inspection is required prior to pouring of the concrete. Purpose of inspection is to check that soil below footings is of the type that is adequate to support loads without excessive settlement.
Alternatives:	Can use a wood or masonry block foundation.
Services by:	Concrete contractor.

6—Foundation Walls and Pier Footings

Completion Time—5 Days
Percent of Total Cost—4.0

Purpose:	Pour basement and crawl space concrete walls and pour footings for steel columns.
Work Included:	Provide and place formwork and concrete to construct wall and column footings in accordance with design drawings. Also included is furnishing and placing of reinforcing as shown on plans, anchor bolts at top of wall, window wells, window frames and glass, and stripping of formwork.
Furnishing Material:	Material provided by concrete contractor.
Inspections:	Inspection is required before pouring concrete. Inspector checks to see if formwork is constructed properly and in the correct location. Inspector also checks whether concrete reinforcing steel is properly placed.
Alternatives:	Can use wood or masonry block foundation.
Services by:	Concrete contractor.

Figure 2-3 Formwork for concrete foundation wall

7—Foundation Drain

Completion Time—1 Day
Percent of Total Cost—0.5

Purpose: Water pressure behind the wall might result in significant lateral pressure on the foundation wall and uplift of the basement slab. The water pressure is reduced by installing a foundation drainage system.

Work Included: Place 4 inch diameter perforated plastic pipe around perimeter foundation at level of footing. Twelve inches of stone is placed on top of pipe to facilitate good drainage characteristics around the area of the pipe.

Furnishing Material: Material provided by concrete contractor.

Inspections: The drain system must be inspected prior to being buried.

Alternatives: Might not be needed in areas that do not have high water tables.

Services by: Concrete contractor.

Figure 2-4 Foundation drain system prior to backfilling

8—Damp Proof Walls

Completion Time—1/2 Day
Percent of Total Cost—0.2

Purpose:	Application of a protective coating on foundation wall that is exposed to the soil. This assists the wall in resisting the elements and provide damp proofing properties.
Work Included:	Labor and material for spraying on two coats of the asphalt compound to the walls.
Furnishing Material:	Material provided by concrete contractor.
Inspections:	The damp proofing system must be inspected before backfilling.
Alternatives:	Might not be needed in areas that do not have high water tables.
Services by:	Concrete contractor.

Figure 2-5 Damp proofing of walls

9—Brace Walls and Backfill

Completion Time—1 Day
Percent of Total Cost—0.4

Purpose:

Walls must be braced since the wood frame of the house and the basement floor have not been installed at this point in the project. The wood frame and basement floor provide lateral support for the concrete wall at the top and bottom, respectively. Prior to their installation, the walls must have temporary bracing, since the wall is unstable. After the wall is braced, the excavation around the foundations is backfilled to bring the soil up to the bottom of the first floor.

Work Included:

Wood bracing sufficient to resist the lateral pressure exerted by the backfill is provided. Equipment and labor to backfill the excavation is provided under this work item.

Furnishing Material:

Material provided by general contractor.

Inspections:

None.

Alternatives:

Work is only required if a basement is to be constructed.

Services by:

Rough framing carpenters.

10—Set Steel Column and Beams

Completion Time—1 Day
Percent of Total Cost—0.5

Purpose:	The wood joists are not capable of spanning the entire length of the basement. For this reason, steel beams supported on steel pipe columns are needed near midspan.
Work Included:	This work item includes the labor needed to install the steel pipe columns and the installation of the steel beams, including the connections. Care must be taken to install the steel beams at the correct elevation.
Furnishing Material:	The steel members are furnished by the general contractor. The steel should be ordered from the fabricator one week prior to the installation date.
Inspections:	None required.
Alternatives:	A wood post and beam system can be substituted for the steel system.
Services by:	Rough framing carpenters.

Figure 2-6 Steel pipe column and beams

11 through 14—Wood Framing

Completion Time—10 Days
Percent of Total Cost—17.0

Purpose: Construct structural frame of the house. Includes first story floors and walls and second story floors and walls.

Work Included: Rough framing contractor provides all labor and equipment necessary to construct wood frame of the house. The carpenters also put the sheathing (plywood or similar substitute) over the wood members. This contractor will usually have to supply his own electrical power.

Furnishing Material: The general contractor is responsible for providing all the material for this phase of the work, including the nails. The wood should be delivered on site in portions to prevent theft that may occur if lumber is stored on site for a long time. Material should be ordered one week prior to starting the work.

Inspections: Inspection of the rough framing by the building official is required.

Alternatives: A concrete or masonry structure can be used in lieu of a wood frame, but this is highly unusual.

Services by: Rough framing carpenters.

Figure 2-7 Wood framing

15—Roof Framing

Completion Time—4 Days
Percent of Total Cost—5.0

Purpose: Construct the structural components of the roof.

Work Included: Work either includes placement and attachment of trusses constructed off-site, or construction and attachment at the jobsite. This item also includes attachment of wood sheathing (plywood) over the truss members.

Furnishing Material: Material is provided by the general contractor. General contractor orders material with other framing lumber or orders prefabricated trusses from a truss manufacturer.

Inspections: Inspection is included with wood framing inspection.

Alternatives: None.

Services by: Rough framing carpenters.

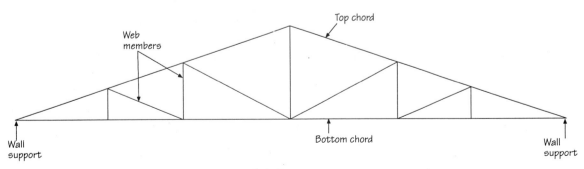

Figure 2-8 Truss components

16—Rough Plumbing

Completion Time—3 Days
Percent of Total Cost—5.0

Purpose:	Provide house with rough plumbing that will be located in places inaccessible after drywall is installed.
Work Included:	Furnish material and perform necessary labor to provide rough plumbing. This includes placement of all clean water and waste water pipes. This work should also include installation of sump pump system, water heater, water softener, as well as all pipe venting.
Furnishing Material:	Material provided by plumbing contractor.
Inspections:	Inspection required before plumbing is covered by insulation.
Alternatives:	None.
Services by:	Plumbing contractor.

17—Heating and Air Conditioning

Completion Time—3 Days
Percent of Total Cost—3.5

Purpose:	Provide house with heating and air conditioning capabilities.
Work Included:	Furnish and install all sheet metal duct work for flow of heated and cooled air. Also, furnish and install furnace, fan, damper equipment, chimney exhaust, and humidifier. Note that the air conditioning unit is not installed at this point. It is recommended that the air-conditioning unit be installed at time of occupancy to avoid theft of the unit.
Furnishing Material:	Material provided by HVAC contractor.
Inspections:	Usually none required.
Alternatives:	Some houses may not need heating or cooling capabilities depending on their geographical location.
Services by:	HVAC contractor.

18—Rough Electric and Telephone Wiring

Completion Time—3 Days
Percent of Total Cost—5.0

Purpose:	Provide house with rough electric and telephone lines.
Work Included:	Furnish material and perform necessary labor to provide rough electric. Cost includes labor to install light fixtures, but not the cost of the fixtures. Also includes all labor and material for telephone lines.
Furnishing Material:	All material for rough electric and phone lines is provided by electrical contractor.
Inspections:	Inspection required before wiring is covered by wall insulation.
Alternatives:	None.
Services by:	Electrical contractor.

19—Windows and Exterior Doors

Completion Time—2 Days
Percent of Total Cost—4.0

Purpose:	Provide house with all exterior doors and windows.
Work Included:	Furnish all doors and windows. Labor costs are usually included in wood framing costs (see Item 11).
Furnishing Material:	Materials are provided by general contractor. Material should be ordered at least three to four weeks in advance.
Inspections:	None.
Alternatives:	None.
Services by:	Rough framing carpenters.

20—Roofing

Completion Time—3 Days
Percent of Total Cost—3.0

Purpose:	Attach roofing material on top of roof plywood. A payment is released to the contractor once the structure is at this stage.
Work Included:	Work includes all labor and materials necessary to roof the structure. This includes roof vents, roof felt, shingles, and flashing.
Furnishing Material:	All material is provided by roofing contractor.
Inspections:	None.
Alternatives:	There are many types of roof shingles including asphalt, wood, and concrete. Also, there is a wide selection of quality for asphalt shingles. The cost for this item was determined assuming asphalt shingles are used.
Services by:	Roofing contractor.

21—Masonry

Completion Time—4 Days
Percent of Total Cost—5.0

Purpose:	Construction of brick facade for house. Also includes fireplace stone.
Work Included:	All material and labor to construct brick facades.
Furnishing Material:	Material is furnished by the masonry contractor.
Inspections:	None required.
Alternatives:	The percent of total cost of the masonry can be as high as 20 percent, depending on the extent of coverage of the exterior of the house. Houses with all exterior walls bricked from ground to roof are not as common as in previous years, since the labor costs are extremely high.
Services by:	Masonry contractor.

22—Stairs and Railings

Completion Time—1 Day
Percent of Total Cost—0.2

Purpose:	Provide prefabricated staircases and railings.
Work Included:	Furnish all stairs, bannisters, and railings. These items are prefabricated and shipped to the job site.
Furnishing Material:	Material provided by general contractor.
Inspections:	None required.
Alternatives:	Stairs can be built on site.
Services by:	Rough framing carpenters or carpenters from prefabricator.

23—Prefabricated Fireplace

Completion Time—2 Days
Percent of Total Cost—0.5

Purpose:	Install prefabricated fireplace.
Work Included:	Furnish and install prefabricated fireplace unit, exhaust chimney, and exhaust pipe, including all necessary hardware.
Furnishing Material:	Prefabricated fireplace contractor furnishes complete prefabricated fireplace system.
Inspections:	Inspection is required in some jurisdictions.
Alternatives:	Build masonry fireplace or omit fireplace.
Services by:	Prefabricated fireplace contractor.

24—Slabs

Completion Time—4 Days
Percent of Total Cost—3.0

Purpose: Pour concrete steps, driveway slab, and basement slab.

Work Included: Spread and compact stone for slab base, and place vapor
                      barrier on top of stone in basement and reinforcing mesh if
                      required, and pour concrete slab. Provide, place, and remove
                      all necessary formwork.

Furnishing Material: All material provided by concrete contractor.

Inspections: Inspection is required prior to pouring concrete.

Alternatives: Slabs can be poured earlier in the construction schedule if
                      desired.

Services by: Concrete contractor.

Figure 2-9 Prepared base for concrete slab

25—Drill Well

Completion Time—2 Days
Percent of Total Cost—6.0

Purpose:	If the property is not situated near an urban area, the house must be provided with potable water from a private well system.
Work Included:	Drilling at specified location to a depth where potable water is reached. Well casing, pump, piping from pump to house, and water test are included.
Furnishing Material:	Material provided by the well drilling contractor.
Inspections:	Water test results must satisfy requirements of state, county, and city.
Alternatives:	Well may not be needed if the municipality has a water supply. In that case, the municipality supply line would be connected to the house.
Services by:	Well drilling contractor.

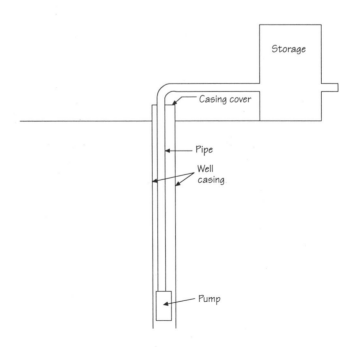

Figure 2-10 Well pumping system

26—Septic System

Completion Time—3 Days
Percent of Total Cost—2.0

Purpose: If the property is not situated near an urban area, the house must be provided with a private waste water treatment facility.

Work Included: Soil investigation and septic design to determine proper layout of waste water treatment facility. Install septic system per design including piping, septic tank, trench excavation, stone, filter paper, and backfill.

Furnishing Material: Material provided by septic installer.

Inspections: Approval required by municipality or county prior to backfilling.

Alternatives: A public waste water treatment system may be available. This would require connecting the waste water effluent pipe to the municipality piping system.

Services by: Septic installer.

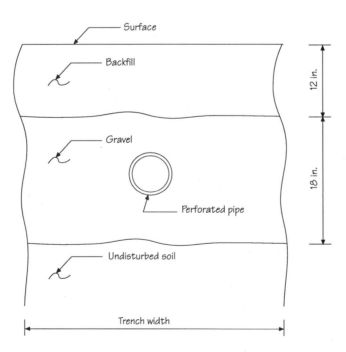

Figure 2-11 Cross section of septic drain pipe

27—Siding

Completion Time—5 Days
Percent of Total Cost—7.0

Purpose:	Provide house with wood siding.
Work Included:	Install wood siding at all locations required by design drawings.
Furnishing Material:	General contractor is responsible for ordering all materials for siding. Material should be ordered one month in advance.
Inspections:	None required.
Alternatives:	Outside of house can be clad fully or partially in masonry, aluminum siding, or vinyl siding.
Services by:	Rough carpenters.

Figure 2-12 Siding installation

28—Garage Doors

Completion Time—1 Day
Percent of Total Cost—1.0

Purpose:	Doors and openers for garage.
Work Included:	Provide and install insulated steel doors with all appurtenances.
Furnishing Material:	Material furnished by garage door installer.
Alternatives:	Can use steel or wood doors.
Services by:	Garage door installer.

29—Gutters

Completion Time—1 Day
Percent of Total Cost—0.5

Purpose:	Water on roof drains down the sloped roof into the gutter that funnels the water to designated locations.
Work Included:	Place aluminum gutters on designated locations of house. Work includes downspouts and all fasteners.
Furnishing Material:	Material provided by gutter installation contractor.
Inspections:	None required.
Alternatives:	Gutters may be omitted.
Services by:	Gutter installation contractor.

30—Insulation

Completion Time—3 Days
Percent of Total Cost—1.5

Purpose:	Insulate inside of exterior walls and ceilings to keep heat or cool air inside the house.
Work Included:	Batt insulation installed in all exterior walls, ceilings, and roof and blown-in insulation into floor of attic space.
Furnishing Material:	Material provided by insulation contractor.
Inspections:	Inspection of all utilities that will be covered by insulation must be performed prior to the start of this work.
Alternatives:	Insulation may not be required in some climates.
Services by:	Insulation contractor.

Figure 2-13 Blown-in attic insulation

31—Drywall

Completion Time—5 Days
Percent of Total Cost—5.5

Purpose:	Provide drywall throughout interior of structure.
Work Included:	Provide drywall attached with screws to walls and ceilings. Drywall will be taped and finished to a quality standard ready for painting.
Furnishing Material:	Material provided by drywall contractor.
Inspections:	None.
Alternatives:	Can use a lath and plaster system.
Services by:	Drywall contractor.

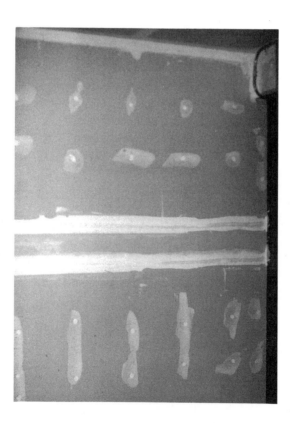

Figure 2-14 Drywall before final finishing

32—Primer Paint

Completion Time—2 Days
Percent of Total Cost—0.5

Purpose:	Place primer paint over drywall.
Work Included:	Place primer coat on all ceilings and walls.
Furnishing Material:	Material provided by paint contractor.
Inspections:	None required.
Alternatives:	None.
Services by:	Paint contractor.

33—Cabinets/Counter Tops

Completion Time—2 Days
Percent of Total Cost—2.5

Purpose:	Provide kitchen and bathroom cabinets. Also provide counter tops for these rooms.
Work Included:	Provide all necessary cabinets and counter tops and all appurtenant items.
Furnishing Material:	Material provided by general contractor.
Inspections:	None required.
Alternatives:	None.
Services by:	Trim carpenters.

34 through 36—Final Painting, Plumbing, and Electric

Completion Time—   Painting–4 Days
                          Plumbing–1 Day
                          Electric–1 Day

Percent of Total Cost—   Painting–2.0
                          Plumbing–0.0
                          Electric–1.0

Purpose: Finish all painting, install plumbing fixtures, and electrical fixtures.

Work Included: Provide second coat of paint. Install all faucets and handles for sinks and showers. Install electrical switches and covers, as well as light fixtures.

Furnishing Material: Material provided by each contractor.

Inspections: None required.

Alternatives: None.

Services by: Painting contractor, electrical contractor, and plumbing contractor.

37—Wood Trim

Completion Time—4 Days
Percent of Total Cost—4.0

Purpose: Construct decorative wood trim on the interior of the house.

Work Included: Install wood trim around doors and windows. Includes installation of wood railings and wood trim on stairs.

Furnishing Material: Material provided by general contractor. Wood trim should be ordered two weeks before it is required on jobsite.

Inspections: None required.

Alternatives: None.

Services by: Trim carpenter.

38—Flooring

Completion Time—3 Days
Percent of Total Cost—4.0

Purpose:	Provide house with floor coverings.
Work Included:	Provide in all rooms of the house, vinyl flooring, ceramic flooring, or carpet. Also included is wall tile for bathrooms. All subfloor for vinyl and tile as well as padding for carpet should be included in this work item.
Furnishing Material:	Material furnished by tile and carpet installers.
Inspections:	None.
Alternatives:	Floors can be exposed hardwood.
Services by:	Tile and carpet installers.

39—Appliances

Completion Time—1 Day
Percent of Total Cost—2.0

Purpose:	Provide household appliances for house.
Work Included:	Furnish and install all necessary appliances, which may include: Refrigerator Stove Washer Dryer Air conditioning unit
Furnishing Material:	General contractor furnishes appliances.
Inspections:	None.
Alternatives:	Owner may already own needed appliances and may not want air conditioning unit.
Services by:	General contractor installs appliances.

40—Mirrors

Completion Time—1 Day
Percent of Total Cost—0.3

Purpose:	Provide mirrors in bathrooms.
Work Included:	Provide and install mirrors.
Furnishing Material:	Material furnished by mirror installer.
Inspections:	None.
Alternatives:	None.
Services by:	Mirror installer.

41—Finish Trim

Completion Time—3 Days
Percent of Total Cost—1.0

Purpose:	Install trim that could not be attached until vinyl flooring was installed.
Work Included:	Install all base molding at floor level. This is a completion of item 37.
Furnishing Material:	Material furnished by general contractor.
Inspections:	None required.
Alternatives:	None.
Services by:	Trim carpenter.

42—Clean Up

Completion Time—3 Days
Percent of Total Cost—0.4

Purpose:	Clean up house to prepare for move in of homeowner.
Work Included:	Washing windows, wash floors, vacuum carpet, clean sinks, and remove all debris.
Furnishing Material:	None required.
Inspections:	Final walk through of homeowner.
Alternatives:	None.
Services by:	Cleaning service company.

43—Spread Top Soil

Completion Time—2 Days
Percent of Total Cost—0.2

Purpose:	Level ground around house and spread black dirt for soil.
Work Included:	Excavation contractor will spread soil that was stockpiled in Item 3.
Furnishing Material:	None needed.
Inspections:	None.
Alternatives:	None required.
Services by:	Excavation contractor.

Chapter 3

Construction Contracts

The construction contract is signed prior to the start of construction and is given little attention until a disagreement between the parties occurs. At this point, the two parties meticulously review the contract in an effort to support their respective positions. In many cases, the contract is poorly drafted and the parties have to litigate to interpret the contract. A well drafted contract avoids this situation. Addressing potential problems in detail, as well as clearly defining the obligations of both parties reduces the need for litigation, since the parties know their respective duties.

This chapter presents and discusses items that, in most cases, should be addressed in a construction contract between the purchaser and the general contractor. The contract agreement between the general contractor and subcontractor is also briefly discussed. This chapter is not meant to be used in lieu of legal counsel, since it is for illustrative purposes and some clauses conflict.

A wide variety of contract clauses are presented throughout this chapter. The actual wording of the clauses are contained within a rectangle with discussion on that clause following. At the end of the chapter, a check list is provided for assistance in contract review.

Purchaser _____

Address _____

_____

Phone _____

General Contractor _____

Address _____

_____

Phone _____

The general contractor agrees to construct a two-story Georgian style home with basement and three-car garage, subject to the terms and conditions of this contract, at 54 South Drake, Your Town, Your State, 12345, whose legal description is:

Lot 7 in Luigina Estates, being a subdivision in the northeast quarter of the southeast quarter of section 18, Township 40 north, range 6, east of the third principle meridian, according to the plat thereof recorded, January 1, 1990, as document #777777 in your county, your state.

The purchaser agrees to purchase the residence at this address under the terms and conditions of this contract.

The precise identification of the parties to the contract needs to be specified to determine exactly who is responsible for obligations of the contract. Parties that are corporations or agents must be identified as such.

It is customary to include a brief description of the project in the beginning of the contract. This description should state the style of house and any other significant requirements.

The address and legal description is needed for the obvious reason of assuring that the house will be constructed on the correct property. This information is also needed for officially identifying the property. The bank providing the construction loan and the general contractor needs this information for official recording of the mortgage and possibly the contract. Also, in the event the general contractor needs to file a lien against the property for lack of payment, a time-consuming search through the official records to find the legal description can be avoided.

(1) Contract Documents—The following items are hereby included as part of this contract:

- Design drawings: Title: _____

    Prepared by:_____

- Attached Specifications
- Local Building Codes
- Occupational Health and Safety Act (OSHA) Requirements
- Local Fire Codes
- All Applicable Material Codes

In the event contract documents conflict, the order of precedence shall be as listed above.

References are made in the contract to the project drawings and other pertinent documents. The references should include the name of the designer who prepared the design drawings, the official title of the drawings, and the order of precedence in the event the documents conflict. All general codes and standards to be enforced for the project should also be mentioned, including state and federal codes. Failure to specify and follow the appropriate codes may result in a structure that will not receive an occupancy permit until expensive retrofitting is completed.

(2) Property Ownership—The purchaser represents that the property is solely and exclusively theirs. The purchaser further represents that the property is free and clear of all liens and encumbrances unless noted elsewhere in this contract.

In many cases, the general contractor is constructing a home on property owned by the purchaser. The general contractor might consider having a title search performed to determine if there are liens on the property of the purchaser. Existing liens can possibly subordinate future liens of the general contractor. The above contract clause requires the owner to have all lien encumbrances remedied prior to signing of the contract as well as requiring exclusive and sole ownership. Building on property that is not owned by the general contractor or purchaser is risky, since limited recourse against the purchaser is available in the event of a breach of contract.

(3) Taxes and Assessments—The purchaser and general contractor shall each pay their pro-rata share of real estate taxes and assessments for the year of closing at the time of closing. The estimated real estate taxes for the year of proration shall be based on 105% of the previous year. The general contractor shall pay any taxes accrued prior to the year of closing.

The general contractor may own the property on which the house will be built and will sell the land and house as a package. The taxes and assessments on the property in the year of transfer must be prorated to the appropriate parties. It is customary that the general contractor pay the taxes and assessments on the vacant property up until the day of transfer. Taxes and assessments are usually paid in the year following their accrual. The contract should clearly state the party paying previously accrued taxes.

(4) Purchase Price and Payouts:

  Base Amount $ _____

  Extras $ _____

  Total $ _____

 Schedule of Payments:

  (i) At contract signing—10%

  (ii) House under roof—35%

  (iii) Interior ready for paint—35%

  (iv) Completion—20%

The contract should reflect the purchase price of the house. Equally important is the schedule of payouts. It is typical to provide the contractor with 10% of the total cost for the purpose of getting mobilized on the project. The remaining balance is usually given to the contractor in two intermediate payments and one final payment.

(5) Loan Commitment—The purchaser must receive a loan commitment within 30 days from signing of this contract in the amount of $XX,XXX.XX. The purchaser shall make every reasonable effort to obtain the loan commitment and shall pay all usual and customary charges imposed by the lending institution. The purchaser shall provide a copy of the commitment letter to the general contractor within the 30 days. If the purchaser is unable to obtain financing, the purchaser shall notify the ge-neral contractor within the 30 days. If the purchaser notifies the general contractor that a loan commitment cannot be obtained, the general contractor has the option to obtain a loan commitment on behalf of the purchaser. All associated fees will be paid for by the purchaser. Failure of the general contractor to exercise this option makes this contract null and void.

This clause requires the purchaser to make a good faith effort to obtain the necessary financing. An option in the clause allows the general contractor to find a loan for the purchaser, if the purchaser is unsuccessful. In the event neither party can get financing for the purchaser, the contract is rescinded. It is clarified in this part of the contract that regardless who obtains the financing, the purchaser must pay all of the associated fees.

> (6) Permits—The purchaser shall pay all fees for any and all permits required for the project, including the building permit. In the event that the design drawings are not acceptable for the obtaining of a building permit, it shall be the purchaser's responsibility to have the drawings modified as necessary at no cost to the general contractor. Other permits include, but are not limited to, building, water, sewer, septic, well, and site grading. The general contractor shall be responsible for obtaining the permits.

This contract clause is dependent on who is providing the design drawings. If the purchaser had the drawings prepared by an independent designer, the purchaser is responsible for assuring the plans are acceptable to the building department. Plans prepared through the general contractor usually shifts the burden to the general contractor.

The above clause clarifies which party pays for all of the necessary permits. However, it places the burden on the general contractor to obtain the permits.

> (7) Testing—The purchaser is responsible for obtaining, as well as paying all fees for any and all testing needed for construction of the residence. Testing includes, but is not limited to, soil strength analysis, permeability tests, radon testing, and water quality testing. The general contractor shall arrange for all testing with the exception of the soil test, which shall be arranged by the owner with an analysis received prior to the start of construction.

Testing, with the exception of soil tests, is required during various phases of construction. It is arranged for by the general contractor, but paid by the purchaser. The payment and arrangement for the soil tests depends on initial ownership of the property. When the general contractor is selling the property and structure as a package, the soil testing is the general contractor's responsibility. If the purchaser owns the property, the soil test is usually arranged and paid for by the purchaser.

> (8) Time of Construction—The structure shall be completed no later than six months from the signing of the contract. The six month period shall be extended for circumstances beyond the control of the general contractor such as, but not limited to, labor strikes, material availability, war, natural disasters, and other unforeseen acts of God. The six month period shall be extended by the length of time of the interruption. The general contractor agrees to pay $100 per day for every day beyond the six months that the structure is not completed. This $100 is compensation for inconveniences and living costs for the purchaser. The purchaser agrees to close on the house, pay the full purchase price, and receive ownership no later than 14 days after completion.

This clause commits the general contractor to a completion date, but still allows for extension of this deadline in the event of circumstances beyond their control. A fine is imposed if the project is not completed on time. These types of fines are called liquidated damages. The wording of this clause must be chosen with care. Liquidated damage clauses are usually limited to the amount of money the purchaser loses because of a late completion date. Unusually large liquated damage clauses have the appearance of being a penalty rather than reimbursement costs. Judges will not enforce these penalties, since it is an encroachment on the function of the court.

> (9) Quality of Work—All work performed by the general contractor and subcontractors hired by the general contractor, shall be in accordance with good construction practice, drawings, and specifications. The work shall meet all applicable building codes and be performed in a quality manner. All work shall be free of defects and faults.

It would seem that it is implied that the general contractor must provide quality work to satisfy their end of the contract. Nevertheless, it is usually expressly stated that the work must meet certain standards.

> (10) Materials—The contractor is required to furnish all material necessary to complete the house. The contractor is the owner of the material, and is therefore responsible for the material until it is installed. The material furnished shall be of good quality, new and at least equal to the quality of the standard in the industry for the respective type of product. The general contractor shall be allowed to substitute material when the specified material is no longer available or is not feasible to purchase. The purchaser shall be notified of such changes and be allowed a reasonable objection to the substitute.

This clause prevents the contractor from using substandard or previously used materials. It also allows the contractor to use substitute materials if the materials are unavailable, or if the cost of a particular item makes it unreasonable to use. The purchaser can object to the change in materials if the objection is well founded. The material is the property of the contractor until the material is installed. This identifies ownership for insurance coverage purposes.

(11) Change Orders—Any changes, deviation, or additional work shall only be performed after a change order has been received by the general contractor. The change order shall contain a description of the work, change in price, and the signature of the purchaser. The change order shall be signed by the general contractor if the general contractor agrees to perform the work at the stipulated price. The general contractor shall not have the right to refuse any reasonable change order. In the case where it is not possible to wait for a properly processed change order, the general contractor shall notify the purchaser of the additional work and price for the change in work. The owner shall provide a verbal response and then provide a properly processed change order within four working days.

Changes from the original contract have the potential to result in big problems. Additional work is not always welcomed by the general contractor because it requires modification of the construction schedule. Also, the purchaser may be skeptical of change order costs, since the owner cannot compare them to other bids. For this reason, changes must be in writing to document the transaction. Such changes can result in litigation nightmares and can best be avoided by well prepared and reviewed design drawings and specifications.

Under some circumstances, there is not enough time to wait for a properly processed change order. For example, the general contractor determines after construction of the roof that the design is inadequate and the roof is unstable. It would be foolish to wait for the processing of a change order while the roof is in an unsafe condition. Performing work without a change order based on a verbal agreement is risky for both parties, but is a necessity in the construction industry.

(12) Purchaser Insurance—The purchaser shall obtain and pay for fire insurance with extended coverage and builders risk insurance for joint coverage of the purchaser, general contractor, and subcontractors. The insurance shall be in the amount of the contract price of the structure excluding concrete work items. The purchaser shall furnish evidence of this insurance to the general contractor and it shall be obtained under the condition that the insurance cannot be cancelled without 30 days notice to the general contractor.

The purchaser is responsible for insuring any and all of the structure. Any material that is not attached to the structure is not the property of or is it insured by the purchaser. The purchaser should also obtain liability insurance, if the purchaser owns the property. All other insurance is typically the responsibility of the general contractor. The amount of coverage does not need to be to the full value of the contract, since concrete items are rarely damaged by fire, theft, or other catastrophes.

---

(13) General Contractor—The general contractor shall obtain and pay for Comprehensive General Liability Insurance and all other insurance required by law. These insurance requirements include, but are not limited to:

- Workman's Compensation

- General Liability—Personal Injury

- General Liability—Property Damage

The general contractor shall provide statutory amounts for insurance required by law and $1,000,000 each person and $2,000,000 each occurrence for general liability insurance. Prior to the start of construction, the general contractor shall provide evidence of the required insurance with a provision that the policy cannot be cancelled without 30 days notice to the purchaser. The policy must cover all material prior to attachment to the structure . All subcontractors employed by the general contractor must maintain insurance in the same amount as required of the general contractor.

Due to the hazardous nature of construction, the general contractor and the subcontractors should not be allowed on the work site without proper insurance. Proper insurance is needed to protect against injuries to workers and others on the job site, to cover any damaged property, and for workmen's compensation. As discussed in the previous item, the purchaser is responsible for insuring the structure and anything that is attached to it. The general contractor is responsible for insuring the material from the time it is delivered until it is installed.

(14a) Warranty provided by General Contractor
Alternative 1—The general contractor warrants all work against defects. The general contractor agrees to correct any defective work at no cost to the purchaser. This warranty is in effect only for one year from the date of completion or date of possession, whichever comes first.

(14b) Warranty provided by General Contractor
Alternative 2—The general contractor warrants all work against defects in accordance with the ABC warranty. All disputes between the purchaser and general contractor shall be resolved according to the procedures outlined in the ABC warranty. Purchaser and general contractor agree to abide by any arbitration decision utilizing the ABC warranty procedure.

Two alternatives are presented for warranties provided by the general contractor. The first alternative is a typical one year warranty provided by the general contractor. The general contractor agrees to repair any defects reported in this time frame. It is important to clarify when the one year period starts.

The second alternative is a warranty backed by a third party. Similar to the previous alternative, the general contractor agrees to repair any defects in the one year period. If the general contractor fails or refuses to make the necessary repairs, the purchaser can make a claim against the third party that backed the warranty. This third party will review and pay for repairs for legitimate claims.

(15) Warranties Provided by Statute—The purchaser is limited to warranties provided by the general contractor as discussed above. The general contractor makes no other warranties, express or implied, including, but not limited to, the implied warranty of habitability, merchantability, or fitness for a particular purpose.

Some states have enacted legislation that protects the purchasers of new homes as well as consumers in general. These laws require that a general contractor provide a certain level of quality as well as a limited warranty. Examples of warranties provided by law are presented and discussed below.

- Implied Warranty of Merchantability—This warranty provided in the Uniform Commercial Code states that a seller of goods must provide goods that meet the minimum level of quality established in the industry.

- Implied Warranty of Fitness for a Particular Purpose—This warranty, also provided by the Uniform Commercial Code, states that if the general contractor knows the purpose the structure is being used for, and the purchaser is relying on the skills of the general contractor to select and furnish suitable goods, there is an implied warranty that the general contractor shall provide goods for that purpose.

- Implied Warranty of Habitability—Several decades ago, the legal system would employ the Doctrine of Caveat Emptor. This doctrine is better known as "buyer beware." This line of thinking basically takes the position that when the purchaser buys the house, all defects were now the problem of the purchaser. Caveat Emptor is no longer followed. The courts' position is now that the general contractor provides an implied warranty that the house is in a habitable condition.

In the absence of a written warranty, the three implied warranties can be utilized by the purchaser. The contract clause presented in Item 15 expressly denies the purchaser the use of these clauses. This limits the purchaser to the general warranties of Item 14.

(16) Safety—The general contractor shall be solely responsible for providing a safe work site. The work site shall be safe for workers and all persons. The general contractor shall be responsible for providing a safe work site and initiating safety programs for all subcontractors. Precautions shall be taken to protect all property. The general contractor is responsible for meeting all federal and state safety requirements, including compliance with OSHA regulations.

Safety is one of the most important issues in construction. A general contractor has a moral obligation to ensure the safety of the worker as well as a financial interest to avoid unnecessary costs and time loss. Since the purchaser is not regularly on the job site, it usually is the general contractor's responsibility to provide a safe working environment. Unfortunately, this responsibility brings a considerable amount of liability with it. The general contractor should work diligently to have a safe work site.

(17) Subcontractors—The general contractor is responsible for ensuring that all work performed by subcontractors is performed in an acceptable manner and in accordance with the contract, design drawings, and specifications. The general contractor shall not use any subcontractors who are not fully insured as required by this contract or any subcontractor reasonably objected to by the purchaser. The contractor shall not be required to hire any specific subcontractor unless expressly specified in this contract.

This clause states with specificity that the general contractor is responsible for all work even if the work is performed by subcontractors. Also, an option is available to the purchaser for refusal of a subcontractor if it is reasonable.

(18) Destruction of Premises—In the event the structure is destroyed before the purchaser has taken ownership, the general contractor shall choose one of the following options.

Within two weeks of the date of occurrence of the destruction, the general contractor shall either:

(i) Rebuild and finish the structure within 180 calendar days using the insurance proceeds, or

(ii) Void the contract, receive the insurance proceeds, and return deposit to purchaser.

The structure can be destroyed by a variety of natural disasters as well as acts of vandalism. When such unfortunate events occur, the contract must leave no ambiguities as to the options of the parties. In the sample text above, the general contractor has two weeks to decide how to proceed. If the general contractor decides to repair and finish the structure, the extra work is paid for by the insurance settlement. Should the general contractor decide not to repair the structure, the general contractor receives the insurance proceeds for the work completed and damaged, but must return the initial deposit. This contract clause is dependent on the premises being properly insured during construction.

(19) Utilities—The general contractor shall be responsible for arranging for permanent utility connections to the residence as well as for the cost incurred to provide utility service to the residence. All costs incurred for utility services prior to the closing shall be paid for by the general contractor. The general contractor shall notify all utility companies prior to any digging to avoid damage to buried service lines.

Arrangements for the cost of providing utility services are paid for by the general contractor. The contract must specify which party is to pay for the cost of utilities (i.e., heat, electricity for carpenters, etc.) during construction.

(20) Clean Up—The general contractor shall keep the construction site in a neat and clean condition during the life of this contract regardless if the work is being performed by the general contractor or a subcontractor. All material shall be stored in an orderly, neat, and safe manner. The entire premises, interior and exterior, shall be cleaned immediately prior to closing, including removal of all trash and debris.

This clause avoids the situation of the general contractor finishing the structure and leaving the premises untenable. It is good practice to keep the construction site in an orderly, clean manner while building the house to avoid injuries or the appearance of a substandard quality project.

> (21) Punch List—At the discretion of the general contractor, the purchaser may take possession of the property upon substantial completion of the residence. If the purchaser does take possession a final "punch list" shall be recorded. The punch list shall list specifically what work is needed to achieve final completion of the contract. The contractor agrees to work diligently to complete the punch list items within 30 days of possession. However, the general contractor shall be paid in full regardless of the unfinished items on the punch list.

Near the end of construction, the general contractor usually has some odds and ends that need to be finished. This minor work should not prevent the purchaser from taking ownership. After the structure is substantially completed, a punch list is prepared and the purchaser takes over ownership. A punch list is construction jargon for a list of unfinished work. The contract must clearly state whether money equal to the value of the punch list work should be retained until the work is completed.

> (22) Assignment of Contract—This contract cannot be assigned to any other party without prior approval from the nonassigning party. However, this contract shall be binding upon and be for the benefit of the parties to the agreement, their heirs and successors, and personal representatives.

An assignment is when one party transfers to another party a right or a duty imposed under the terms of a contract. Assignment is typically forbidden in a construction contract because a change in the general contractor or purchaser can lead to problems. The parties will allow the contract to be passed on to heirs should the contracting party become deceased.

> (23) Default—In the event that the contractor fails to carry out the terms of this contract, the purchaser shall make a written notice to the contractor of the lack of progress. If the contractor does not respond in 10 working days, the purchaser may terminate the contract. The general contractor shall pay to the purchaser any additional cost beyond the original contract for completion of the structure. In the event the purchaser fails to make a required payment, the contractor shall make a written protest of the missed payment. If the purchaser does not make payment, the contractor may void the contract and receive payment for all work completed. The nonbreaching party shall be compensated for any losses including attorney's fees for the breach.

The parties must contemplate the possibility that the other party may not perform and will breach the contract. The contract must list the procedure and remedies in case of a default.

> (24) Arbitration—Any controversy or claim arising out of or relating to this contract, or the breach thereof, shall be settled by arbitration in accordance with the Construction Industry Arbitration Rules of the American Arbitration Association and judgment upon the award rendered by the arbitrator or arbitrators may be entered in any court having jurisdiction thereof.

The insertion of this clause into the contract provides for arbitration should a dispute arise. Alternate methods of arbitration or dispute resolution are available in addition to the American Arbitration Association referenced in the clause. Arbitration has many benefits over a court trial. Arbitration is expeditious, economical, and informal. However, arbitration does have drawbacks. First, the arbitrator is the judge of what evidence can be heard and not required to follow the federal and state rules of evidence. In a court trial, the parties can predict fairly well what can be admitted as evidence. This is not the case in arbitration. Another drawback is that there is little chance of the court overturning a bad decision by the arbitrator.

Although there are problems with arbitration, it is nevertheless a very efficient manner of dispute resolution in the construction industry. The parties should consult an attorney, or the applicable rules on arbitration, to determine if this is the method desired to settle disputes.

> (25) Entire Agreement—This agreement constitutes the entire agreement and is a final complete expression of the agreement between the parties. All prior discussions, promises, or representations are merged into this document.

This clause is a warning that any promises made by either party in the past need not be fulfilled unless it is in this contract. This prevents either party from trying to use parol evidence. Parol evidence is evidence of an earlier oral or written expression used to contradict the contract at issue. The above clause eliminates all previous agreements and discussions.

The previously listed 25 are a partial list of the items that should be addressed in a construction contract. Since every project is unique, every contract is also unique. Fig. 3-1 gives a helpful check list for review of construction contracts.

To this point, the discussion has focused on the contract between the purchaser and general contractor. There are also a multitude of contracts between the subcontractors and the general contractor. These contracts are in much less detail than the previously discussed contracts. In fact, many general contractors and subcontractors work with oral contracts. Though this is not recommended, it has been widely used between parties that have worked together in the past and have built up a level of trust.

The contract between the general contractor and subcontractor usually contains a brief description of the work to be performed and includes the following:

- All work to be completed in a workmanlike manner.
- Changes may only be made in writing and will result in extra charges.
- All agreements contingent upon the occurrence of strikes or accidents beyond the subcontractors control.
- Owner to carry necessary insurance including tornado and fire insurance.
- Method of payment and scheduling of payment.
- Length of time within which proposal must be accepted.
- Interest charges on past due accounts.
- Past due accounts will result in a payment of a collection charge by general contractor.

Contract Clause	Is included in contract	Should be omitted from contract	Does not apply	Notes
(1) Contract documents				
(2) Property ownership				
(3) Taxes and assessments				
(4) Purchase price and payouts				
(5) Loan commitment				
(6) Permits				
(7) Testing				
(8) Time of construction				
(9) Quality of work				
(10) Materials				
(11) Change orders				
(12) Purchaser insurance				
(13) General contractor insurance				
(14) Warranty provided by general contractor				
(15) Warranties provided by statute				
(16) Safety				
(17) Subcontractors				
(18) Destruction of premises				
(19) Utilities				
(20) Clean-up				
(21) Punch list				
(22) Assignment of contract				
(23) Default				
(24) Arbitration				
(25) Entire agreement				

Figure 3-1 Construction Contract Checklist

Chapter 4

Building Plans

Building plans provide the bridge for ideas to be built into reality. These drawings are prepared by architects, engineers, or other competent professionals. The designers must develop two-dimensional sketches that represent a three-dimensional structure. Failure to produce quality drawings, failure to follow the building plans, or failure of the owner to understand what is to be constructed turns the project from a dream to a nightmare.

This chapter assists in understanding building plans for residential structures. A simple step-by-step procedure is used that allows the reader to focus on a limited amount of information at a time.

Building plans come in many standard sizes. However, the two most common sizes used are 36 in. wide by 24 in. high and 22 in. wide by 17 in. high. Typically, they are drawn in pencil on a vellum-like paper. This paper is then used in a blueprint copy machine to provide the many copies that are needed for the project.

Building plans for homes have significantly fewer details than plans for commercial buildings or bridges. In some cases, work to be performed by some subcontractors is typically not shown on the home building plans (plumbing, for example). However, it should be remembered that the building plans are not the only documents that are used to construct a house. The specifications and contracts also provide information that explain the project.

Before the construction begins, a thorough understanding of the project is necessary. A recommended procedure for understanding a construction project is:

1) Briefly review drawings, specifications, and contracts.

2) Make a visit to the proposed construction site.

3) Review each page of the building plans. The typical procedure followed by drawing reviewers is to highlight, in yellow, every line on a drawing that is correct and understandable. Highlight, in red, every line that is not clear and resubmit to the designer to clarify.

4) Closely review the specifications.

5) Review corrected building plans and check compliance with specifications.

Drawings 1 through 7 presented at the end of this section represent a full set of building plans for construction of a residential structure. These drawings are:

1—Foundation Plan (explained in Drawings 1A through 1L)

2—First Floor Plan (explained in Drawings 2A through 2N)

3—Second Floor Plan (explained in Drawings 3A through 3H)

4—Sections (explained in Drawings 4A through 4L)

5—Elevations (no explanation needed)

6—Side Elevations (no explanation needed)

7—Details (explained in Drawings 7A through 7F)

A novice might find these drawings intimidating and confusing. Careful reading of this chapter shows that they are in fact logical and easy to understand.

The method used to help the reader understand the drawings is a step-by-step procedure. For example, Drawing 1 is broken into 12 sequential drawings (1A through 1L). Each drawing is accompanied by explanatory text. Each sequential drawing adds more detail than the previous drawing, until the drawing is complete.

There are other methods to construct this house (i.e., wood foundation rather than a concrete foundation). It would be impractical and confusing to cover every possibility, and although some areas of the United States may not use some of the design principles presented here, it should still stimulate critical reader thinking regarding alternate procedures.

The following are some definitions that are helpful in reading this chapter.

Absorption Field—A septic system consists of a number of lines of perforated pipe. The perforated pipe distributes the waste water over a large area of land where it percolates into the soil. The area of land is known as an absorption field.

Base Material—A layer of properly sized stone that is consolidated to serve as the base for a concrete slab. The base material provides adequate drainage characteristics as well as assists in distributing loads to the underlying soil.

Bearing Wall—A wall that supports load, and different than a partition wall, which only supports its own weight.

Bond Breaker—A material placed between adjacent concrete members to prevent the two members from bonding together when the concrete dries.

Brick Veneer—A wall of bricks that is placed directly in front of a structural wall. The brick is decorative and only carries its own weight.

Bridging—Short wood members located between joists, usually placed in the shape of an "X" that assist in spreading concentrated loads on a single joist to the adjacent joists.

Elevation— 1) A term used in drafting to represent the 2-dimensional view of a 3-dimensional object by projecting that object on a vertical plane.

2) The depth below ground or height above ground where a particular object is located.

Footing—The widened piece of concrete located below a concrete wall that assists in distributing the load of the wall over a larger area of soil.

Forms—Wood panels that are used to retain wet concrete until it gains adequate strength. Forms are typically reusable.

Frieze Board—A decorative board placed at the top of a structure that has aluminum or wood siding. The board is located at the wall/roof interface.

Frost Level—The lowest depth below ground at which the air temperature causes the soil to freeze. Items located below this level will not experience heave from freezing soil.

Gas Curb—A step up from the garage floor to the house is to keep the garage elevation below that of the house. This helps reduce the inflow of automobile exhaust into the house.

Ground Fault Indicator (GFI)—A special type of electrical outlet that disables itself if moisture is detected to prevent electrocution to the user.

Header—A structural member that is located above an opening in a wall to carry the loads over the discontinuity. Headers are located above window and door openings.

Joist—A structural component of the floor system. Usually is 2×10s or 2×12s that support the weight of the floor.

Keyway—A void placed into the top of a concrete footing that is later filled by the wall above. The filled void prevents the wall from slipping off of the footing when lateral loads from the retained soil are applied.

Laminated Beam—A beam manufactured by taking thin pieces of wood and gluing them together with the grain in a predetermined pattern. A heat and pressure treatment is then used to permanently bond the material. The beam has excellent structural characteristics.

Lateral Pressure—A force that acts in a horizontal direction. Example of lateral pressures are wind loads, earthquake motion, and pressure caused by soil pushing against a wall.

Moisture Barrier—Impermeable material that is placed in walls and below concrete slabs to prevent moisture from passing into the living space.

Partition Wall—A wall that does not support any loads except for its own weight. A partition wall serves only to divide up an open area into smaller areas.

Pre-engineered Trusses—A term used in residential structures for trusses that are designed and fabricated by a company other than the rough framing carpenters building the house.

Plan View—A term used in drafting to represent the view of a structure that would be seen if one were in the sky looking down on the house.

R-Value—A numerical value that represents insulation performance. A higher R-value represents a better performing insulation.

Riser—The part of a stair step that is vertical.

Scaling—Measuring lengths off a drawing and then increasing them proportionately with a predetermined factor to determine the actual dimension to be used in the field.

Septic System—A wastewater treatment system used in areas where a central wastewater treating facility is not available. A septic system is a two phase system. Phase one is the collection of the sludge in a tank. Phase two is the spreading and permeating of the waste water into the absorption field.

Specifications—Written instruction concerning the requirements for a construction project. A project is built using both the specifications and drawings.

Stud—Vertical members that are part of the wall system.

Sump Pit—Typically, a plastic bucket that is buried flush with the basement slab. Water from the foundation drain flows into this container and then be pumped up and out to a location away from the house.

Tread—The part of a stair step that is horizontal that the foot is placed on.

Welded Wire Fabric—Reinforcing used for concrete slabs to constrain cracks to microscopic widths. The reinforcing consists of steel wires placed in perpendicular directions to form a mesh.

Full Set of Residential Plans

The next seven pages constitute a full and complete set of residential drawings. A review of these seven drawings will be confusing to the novice. Do not be discouraged. The sole purpose of this chapter is to provide a complete understanding of these drawings. Review them quickly on your first pass through the chapter. After you have completed the chapter, return to these pages to review them with your recently gained knowledge.

The drawings that make up the complete set are:

Drawing 1—Foundation Plan

Drawing 2—First Floor Plan

Drawing 3—Second Floor Plan

Drawing 4—Sections

Drawing 5—Elevations

Drawing 6—Side Elevations

Drawing 7—Details

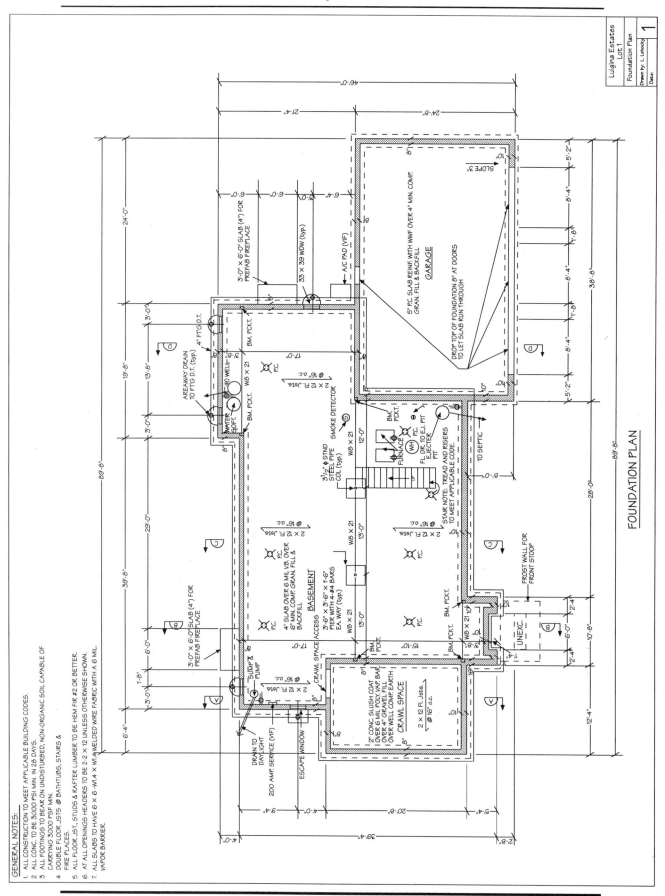

FOUNDATION PLAN

GENERAL NOTES:

1. ALL CONSTRUCTION TO MEET APPLICABLE BUILDING CODES.
2. ALL CONC. TO BE 3000 PSI MIN. IN 28 DAYS.
3. ALL FOOTINGS TO BEAR ON UNDISTURBED, NON-ORGANIC SOIL CAPABLE OF CARRYING 3000 PSF MIN.
4. DOUBLE FLOOR JSTS @ BATHTUBS, STAIRS & FIRE PLACES.
5. ALL FLOOR JST, STUDS & RAFTER LUMBER TO BE HEM FIR #2 OR BETTER.
6. AT ALL OPENINGS HEADERS TO BE 2-2 x 12 UNLESS OTHERWISE SHOWN.
7. ALL SLABS TO HAVE 6 x 6 -W1.4 x W1.4 WELDED WIRE FABRIC WITH A 6 MIL. VAPOR BARRIER.

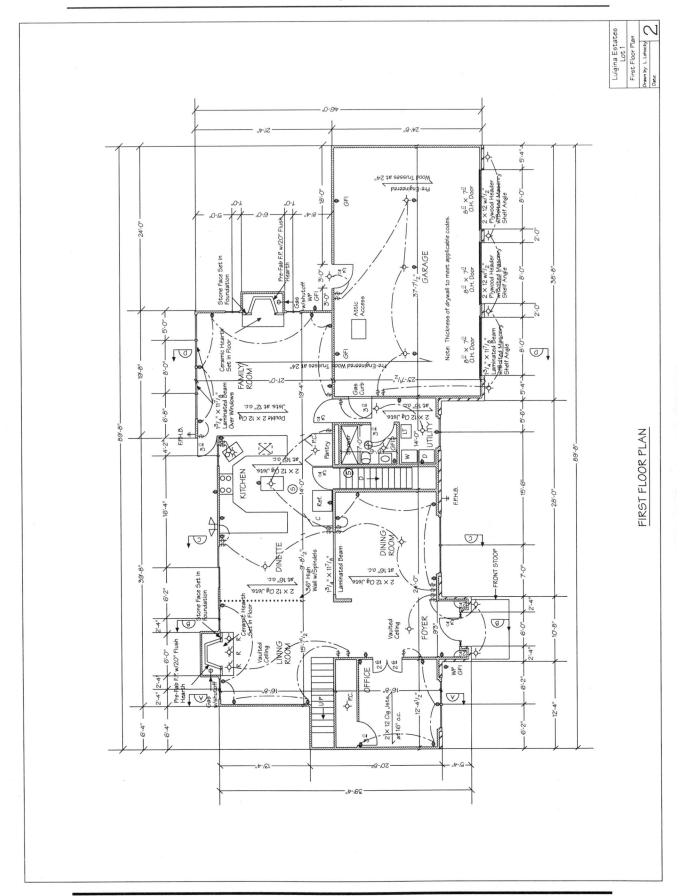

FIRST FLOOR PLAN

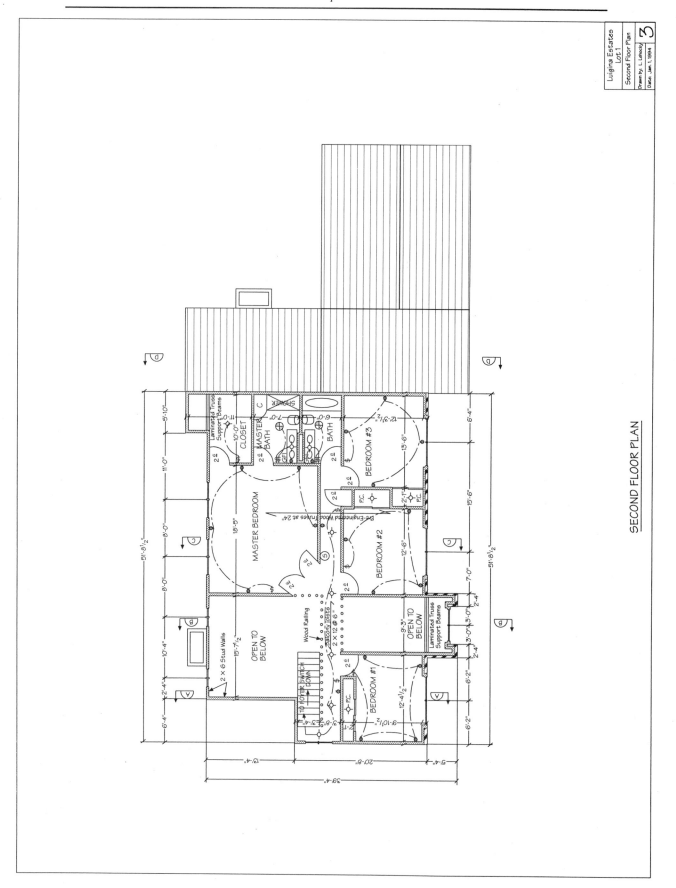

SECOND FLOOR PLAN

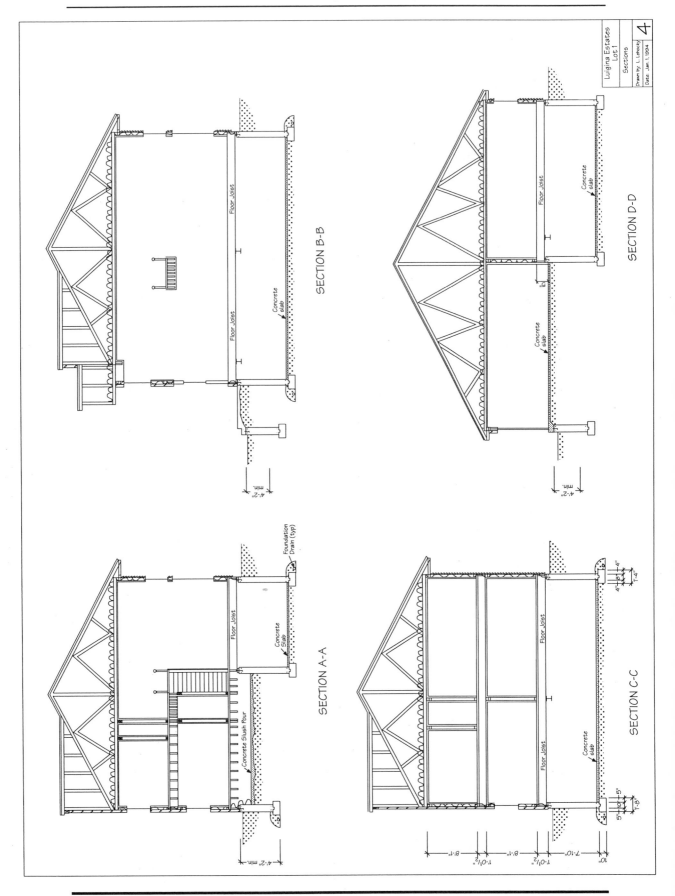

SECTION B-B

SECTION D-D

SECTION A-A

SECTION C-C

Floor Joist

Concrete slab

Concrete Slush Pour

Foundation Drain (typ)

Concrete Slab

4'-2" min.

Luigina Estates
Lot 1
Sections
Drawn by: L. Lehocky
Date: Jan. 1, 1994
4

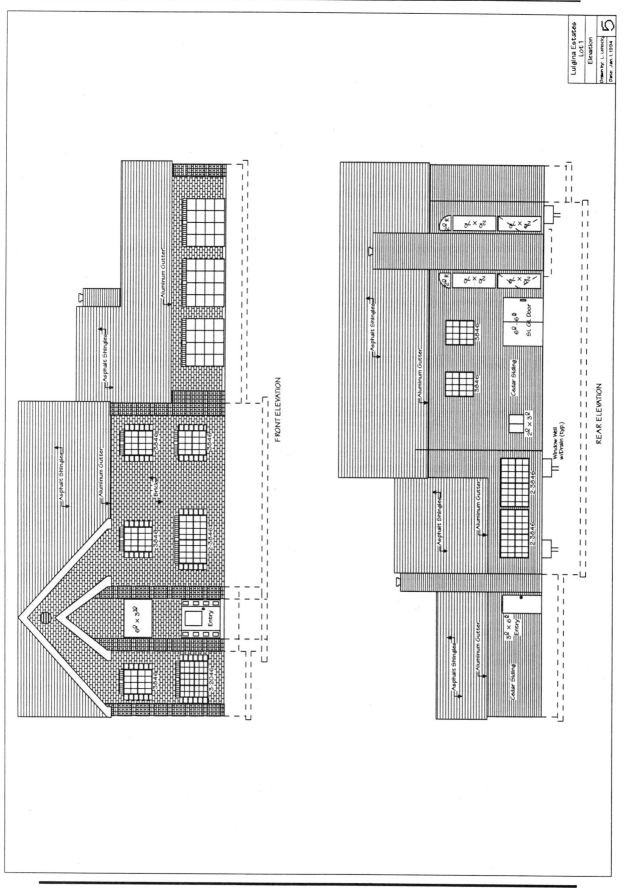

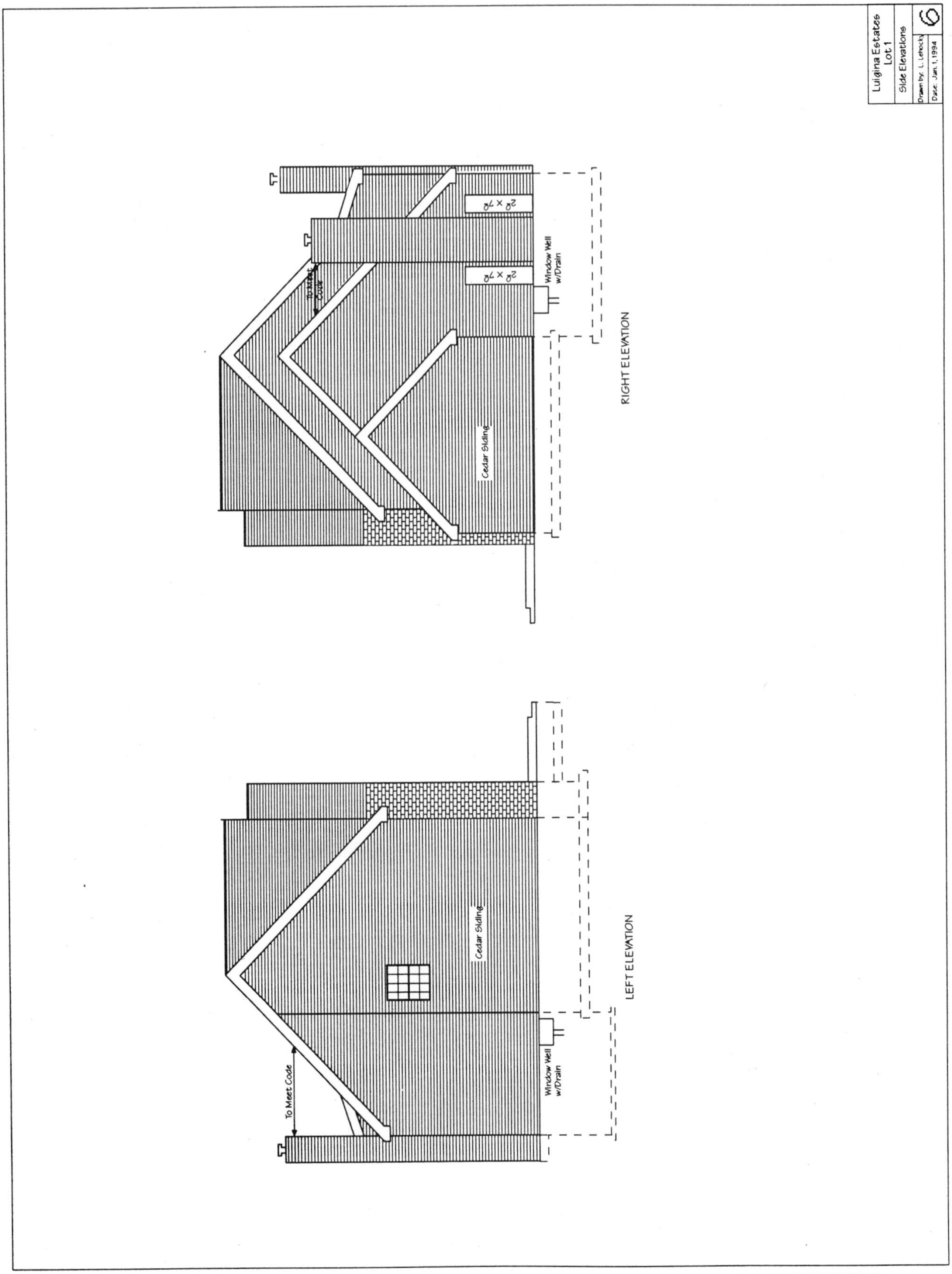

Luigina Estates
Lot 1
Side Elevations
Drawn by: L. Lubecki
Date: Jan 1, 1994

RIGHT ELEVATION

2⁰ × 7²

2⁰ × 7²

Window Well
w/Drain

Cedar Siding

To Meet Code

LEFT ELEVATION

Cedar Siding

Window Well
w/Drain

To Meet Code

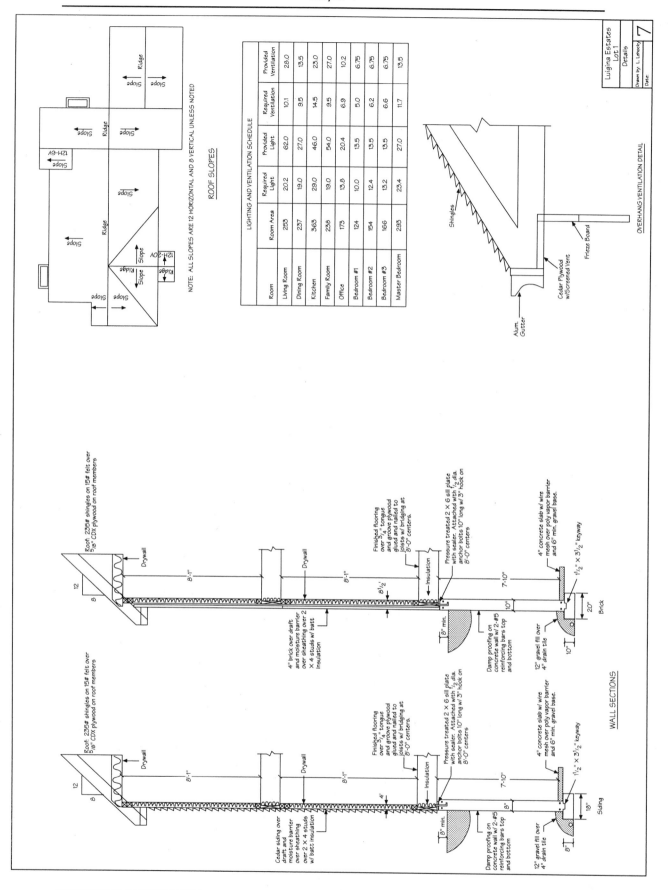

NOTE: ALL SLOPES ARE 12 HORIZONTAL AND 8 VERTICAL UNLESS NOTED

ROOF SLOPES

LIGHTING AND VENTILATION SCHEDULE

Room	Room Area	Required Light	Provided Light	Required Ventilation	Provided Ventilation
Living Room	253	20.2	62.0	10.1	28.0
Dining Room	237	19.0	27.0	9.5	13.5
Kitchen	363	29.0	46.0	14.5	23.0
Family Room	238	19.0	54.0	9.5	27.0
Office	173	13.8	20.4	6.9	10.2
Bedroom #1	124	10.0	13.5	5.0	6.75
Bedroom #2	154	12.4	13.5	6.2	6.75
Bedroom #3	166	13.2	13.5	6.6	6.75
Master Bedroom	293	23.4	27.0	11.7	13.5

OVERHANG VENTILATION DETAIL

WALL SECTIONS

Luigina Estates
Lot 1
Details

7

Drawn by: L. Lohocky
Date:

1—Foundation Plan

Drawing 1A—Foundation Wall Footing

Drawing 1B—Foundation Wall

Drawing 1C—Window Wells, Column Piers, Pre-fabricated Fireplace Concrete Pad

Drawing 1D—Steel Beams and Steel Columns

Drawing 1E—Wood Floor Joists

Drawing 1F—Foundation Drain System

Drawing 1G—Slabs and Stairways

Drawing 1H—Utilities and Section Marks

Drawing 1I—General Notes

Drawing 1J—Electrical

Drawing 1K—Exterior Dimensions

Drawing 1L—Interior Dimensions

Drawing 1A —Foundation Wall Footing

The weight of the house and any other applied loads (i.e., snow loads on the roof, furniture and appliance loads, etc.) must be carried by the structure, transferred to the foundation, and then transferred to the underlying soil. The load on the foundation must not exceed the strength capacity of the soil that the foundation rests on. Settlement of the structure from soil movement will cause cracking in brick walls and the plaster, door jamming, and possibly collapse. This problem is avoided by pouring the wall on a narrow slab of concrete that is slightly wider than the wall, called a footing. The purpose of the footing is to provide increased area at the bottom of the wall to spread the loads and decrease the pressure on the soil. A photograph of a typical footing is shown in Fig. 4-1. A footing is needed at the bottom of each foundation wall.

A footing is constructed by excavating to the proper elevation (determined by the soils engineer) and placing braced wood planks to retain the poured concrete as shown in Fig. 4-2. These wood planks are typically 2 × 10s or 2 × 12s and are called forms. The forms are greased prior to pouring of the concrete so they do not bond to the concrete.

Immediately after the concrete has been poured, a keyway must be formed in the footing. The keyway is a groove in the top of footing that assists in preventing lateral movement of the foundation wall (see Fig. 4-1). A keyway is formed by placing a greased piece of lumber in the concrete. The lumber can be removed as soon as the concrete has set enough to retain its shape. The keyway is typically not shown on the foundation drawing (see drawings 4 and 7).

The depth dimension of the footing is usually the same thickness as the wall. The width of the footing depends on whether the footing is beneath a wall that is supporting masonry. If the side of the house is brick, the footing is typically 20 inch wide. If the side of the house is siding, an 18 inch wide footing is typical. The concrete for the footing is specified to attain a strength of 3000 psi in 28 days. This provides sufficient strength to support the loads and a concrete mix that is easy to place.

Figure 4-1 Concrete footing

Figure 4-2 Footing formwork

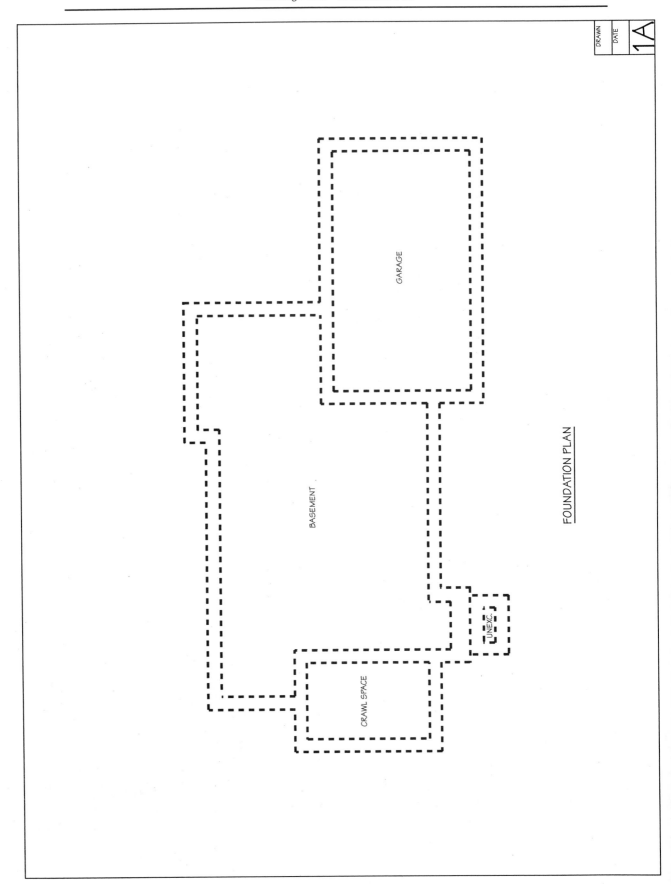

GARAGE

BASEMENT

UNEXC.

CRAWL SPACE

FOUNDATION PLAN

DRAWN

DATE

1A

Drawing 1B—Foundation Wall

The foundation wall must transfer the loads from the structure to the underlying soil, as well as retain the soil behind the basement walls. Foundation walls can be constructed of wood, concrete, or concrete block. The design presented in this publication utilizes concrete as shown in the cross section in Fig. 4-3.

A concrete foundation wall is constructed by pouring concrete between walls constructed of wood, called forms, that hold the concrete while it is setting. After the concrete has set, the wood is removed. If the foundation wall will have windows, or if pipes will protrude through the wall, they are placed directly in the forms so that they will be embedded in the concrete. Fig. 4-4 shows formwork in place and ready for the concrete pour.

Concrete foundation walls are 10 inches thick if they support a wall with a brick face. If wood or aluminum siding is used, the thickness required is only 8 inches. Foundation walls are not reinforced with steel bars with the exception of two lines of horizontal bars at the top and bottom of the wall to prevent the propagation of cracks. The concrete is specified to be 3000 psi in 28 days as was used for the foundation wall footing.

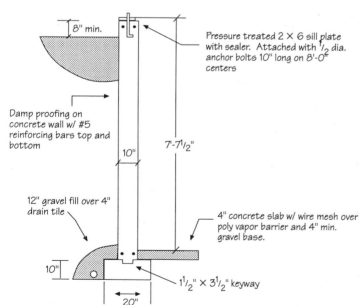

Figure 4-3 Cross section of foundation wall

Figure 4-4 Wall formwork

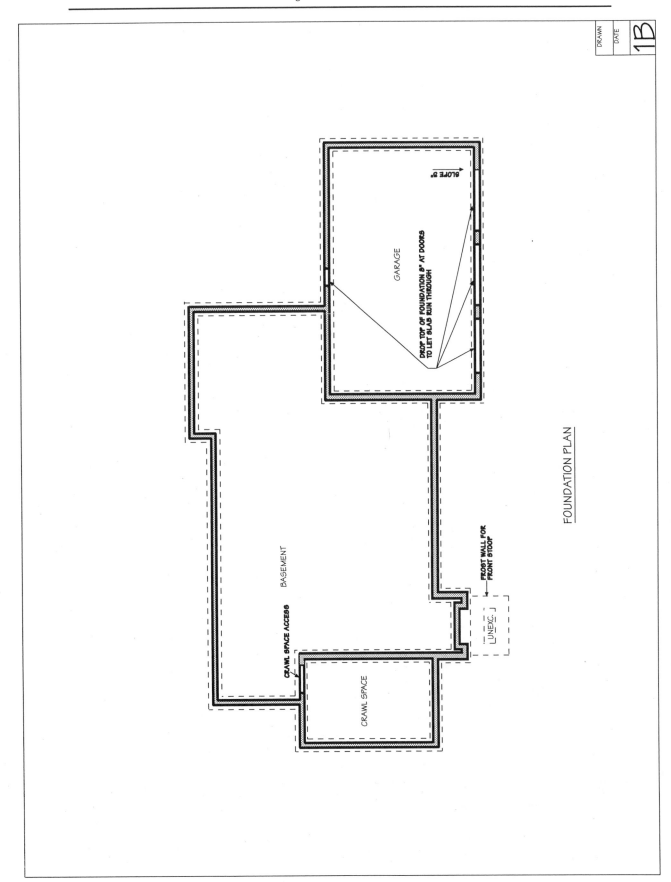

FOUNDATION PLAN

DRAWN

DATE

1B

Drawing 1C—Window Wells, Column Piers, Prefabricated Fireplace Concrete Pad

A basement must have some amount of natural light. This is achieved by placing windows in the foundation walls. Foundation walls are below the ground level and soil must be retained a distance from the wall to allow the exterior light to reach the window. To achieve this, the soil is retained by window wells. Window wells are half circles or rectangles of corrugated steel as shown in Fig. 4-5. The corrugated steel window wells are attached to the foundation wall with bolts.

An emergency exit should be provided for the basement in addition to the interior stairway. The requirement is met by using a basement window well that is also an escape window. The window sill must not be more than 44 inches above the floor and the window must be able to be removed without the use of tools (i.e., a pull chain) to facilitate escape during a fire.

Interior loads on the structure are partially supported by steel pipe columns. These columns have heavy loads requiring them to be supported on concrete piers (Fig. 4-6) to reduce the soil pressure to an acceptable level. The size of these piers is a function of the soil strength and the column loads. The steel pipes are connected by bolts to the concrete piers.

Prefabricated fireplaces are much lighter than the traditional masonry fireplace. These fireplaces require only a concrete pad at ground level for support rather than a foundation. The dimensions of the slab are determined by the size of the fireplace unit to be used. If a traditional fireplace is used, it requires a regular foundation rather than a slab.

Figure 4-5 Window wells

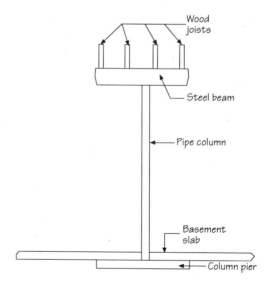

Figure 4-6 Column pier

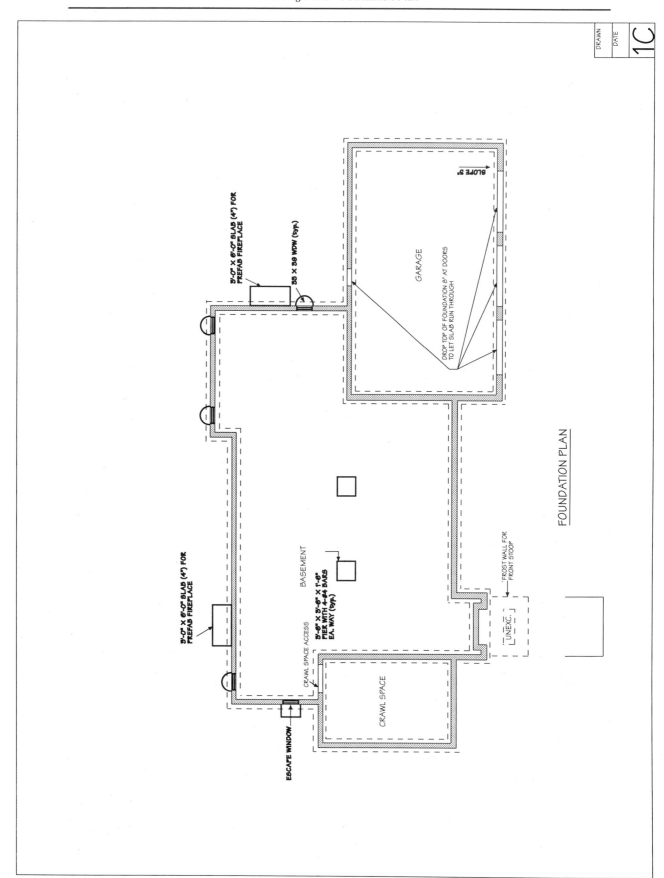

3'-0" X 6'-0" SLAB (4") FOR
PREFAB FIREPLACE

55 × 58 WDW (typ.)

3'-0" X 6'-0" SLAB (4") FOR
PREFAB FIREPLACE

CRAWL SPACE ACCESS

BASEMENT

3'-6" X 3'-6" X 1'-0"
PIER WITH 4—#4 BARS
EA. WAY (typ.)

ESCAPE WINDOW

CRAWL SPACE

GARAGE

SLOPE 3"

DROP TOP OF FOUNDATION 8" AT DOORS
TO LET SLAB RUN THROUGH

UNEXC.

FROST WALL FOR
FRONT STOOP

FOUNDATION PLAN

DRAWN

DATE

1C

Drawing 1D—Steel Beams and Steel Columns

Since the wood joists cannot span the entire width of the house, steel beams are used at midspan to reduce the span length of the wood joists. The steel beams must be of sufficient size that they will not be overstressed and deflection constraints will not be exceeded. The beams must be designed by a professional trained in structural analysis.

The steel beams are designated on this drawing as W8 × 21. The W designation denotes that the steel beam is a wide flange beam (see Fig. 4-7). The number 8 is the approximate height of the beam in inches. The number 21 is the weight in pounds per foot of the beam. Steel beams are sold by weight and it is economical to use the lightest beam that satisfies the design requirements.

The steel beams are supported by the concrete foundation walls and steel pipe columns. Pockets are formed in the top of the foundation walls for the steel beams to rest in (see Fig. 4-8). These pockets are formed by placing styrofoam blocks in the top of the concrete walls immediately after pouring and removing them after the concrete has set.

Diameter of steel columns used in the house are either 3½ or 4 inches. These pipes are filled with concrete and are bolted to the concrete pier at the base and to the steel beams at the top (Fig. 4-9).

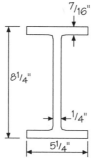

Figure 4-7 Dimensions of typical steel beam (W8 × 21)

Figure 4-8 Pockets for steel beams

Figure 4-9 Connection of steel pipe column to steel beam

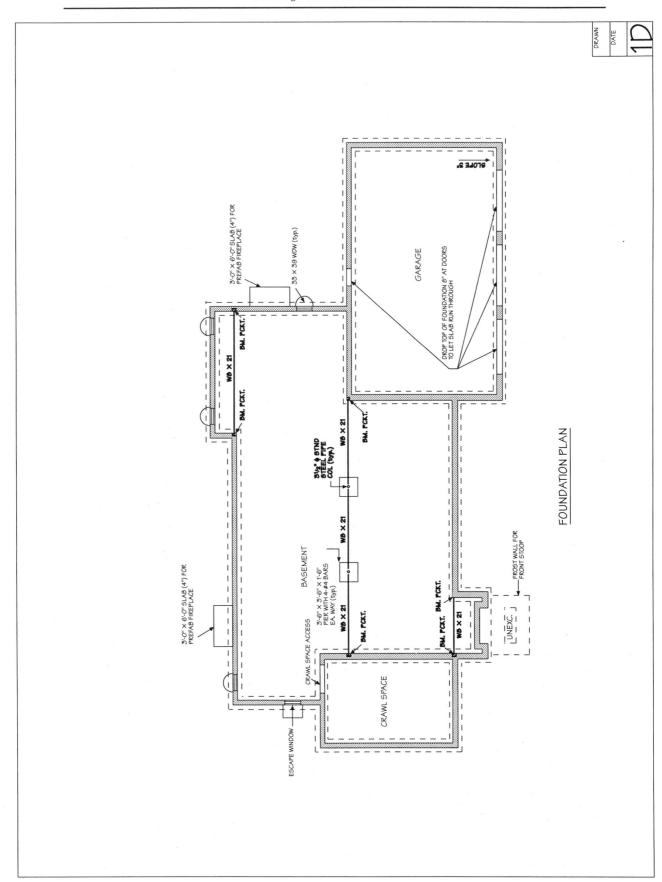

FOUNDATION PLAN

3'-0" X 6'-0" SLAB (4") FOR
PREFAB FIREPLACE

3.5 X 3.9 WDW (typ.)

BM. PCKT.

W8 X 21

BM. PCKT.

GARAGE

DROP TOP OF FOUNDATION 8" AT DOORS
TO LET SLAB RUN THROUGH

SLOPE 3"

W8 X 21

BM. PCKT.

5½" ∅ STND
STEEL PIPE
COL (typ.)

W8 X 21

BM. PCKT.

BM. PCKT.

W8 X 21

FROST WALL FOR
FRONT STOOP

UNEXC.

3'-0" X 6'-0" SLAB (4") FOR
PREFAB FIREPLACE

BASEMENT

CRAWL SPACE ACCESS

3'-6" X 3'-6" X 1'-6"
PIER WITH 4-#4 BARS
EA. WAY (typ.)

W8 X 21

BM. PCKT.

CRAWL SPACE

ESCAPE WINDOW

DRAWN

DATE

1D

Drawing 1E—Wood Floor Joists

Floor loads are carried by the plywood and joist floor system. Wood joists are typically 2 × 10, 2 × 12, or 2 × 14. Note that the size designation for lumber is different than the actual member dimensions. For instance, a 2 × 10 joist is actually 1¾" wide by 9½" deep, not 2" by 10". Joists are evenly spaced at 12 or 16 inches from adjacent joists.

Many species of lumber are available for joists. Each type of lumber has different characteristics with respect to deflections and load carrying capacity. Design aids are available that can assist the designer in selecting the proper joist size if the species of lumber is known. Fig. 4-10 lists types of wood joists and typical span lengths.

Floor joists are supported on the top of the foundation, steel beams, or bearing walls. The joists shown on the drawings span in the direction of the arrowheads. Fig. 4-11 shows the underside of a floor system.

	Span length with 12" centers for 2 × 12	Span length with 16" centers for 2 × 12
Hem Fir	21'-0"	19'-1"
Northern Pine	20'-6"	18'-7"
Aspen	18'-9"	17'-0"
Northern White Cedar	16'-8"	15'-2"
Cottonwood	19'-4"	17'-7"

*40 psf live load

Figure 4-10 Span lengths for various types of lumber

Figure 4-11 Underside of joist floor system

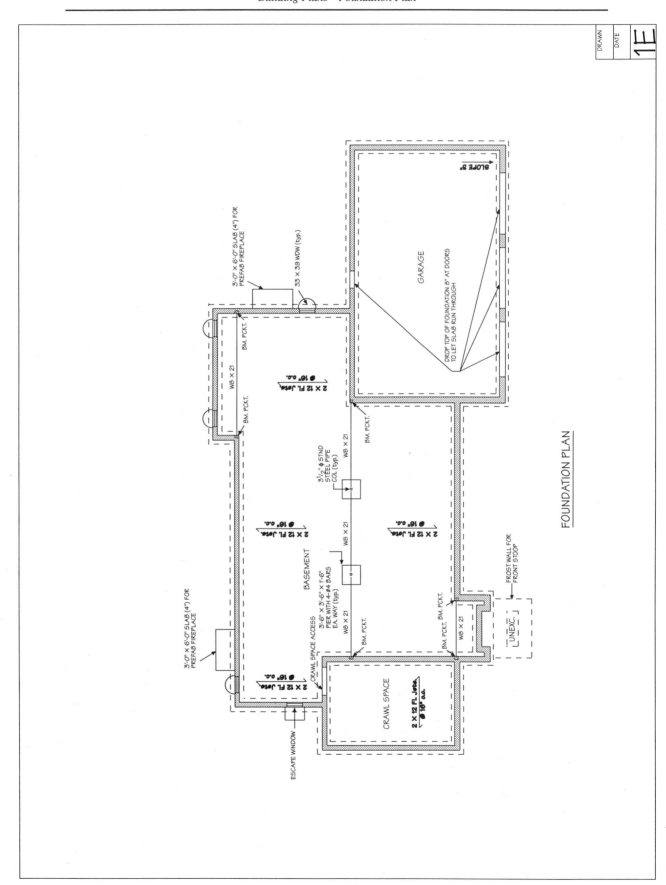

FOUNDATION PLAN

Drawing 1F—Foundation Drain System

The location of the natural ground water level can be a foot to many feet below ground depending on the topology of the land. This natural water causes a significant increase in the lateral pressure on the foundation wall. It can also result in uplift of the basement slab. These pressures can be reduced by the use of a foundation drain system.

A foundation drain system is constructed of a perforated continuous plastic pipe around the circumference of the foundation next to the footing, Fig. 4-12. The water in the soil near the wall drains into the gravel that is placed around the pipe and then into the holes in the pipe. Once the water is in the pipes, it flows into the sump pit in the house and pumped up and out of the basement to a location away from the foundation.

Rain water can accumulate in the window wells and eventually leak through the basement windows. This problem can be avoided by placing a drain in the window well that is connected to the foundation drain, Fig. 4-13.

Figure 4-12 Perforated plastic pipe adjacent to wall footing shown prior to backfilling

Figure 4-13 Window well drain shown prior to backfilling

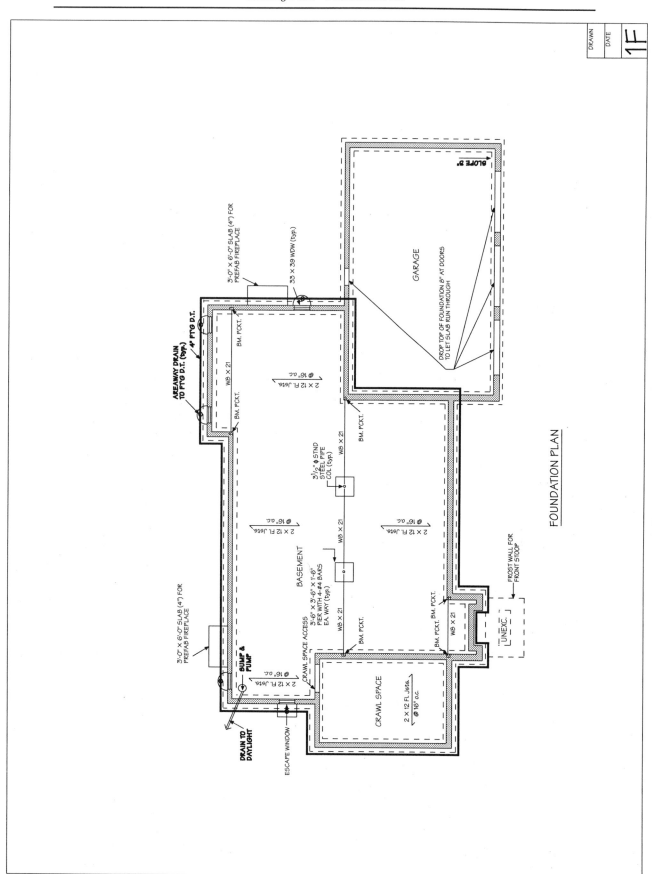

FOUNDATION PLAN

Drawing 1G—Slabs and Stairway

Finished concrete slabs are required for the basement floor, garage floor, and as a base for the air conditioning unit. A slab without the top finished is used in the crawl space.

Quality construction requires that the slab be constructed on top of a compacted granular fill. The compaction reduces the possibility of the soil sinking from the weight of the slab. The requirement of the fill being granular allows for good water drainage characteristics to prevent moisture problems.

The slab thickness is usually 4 inches, but a slightly thicker slab is used in the garage, usually 5 or 6 inches. Before the slab is poured, it is advisable to place a moisture barrier below the slab. A moisture barrier is typically provided by using sheets of plastic placed below the slab. This plastic prevents the migration of moisture from below the slab into the basement. Slabs are usually not reinforced with steel bars, however, a welded wire fabric (WWF) is often used in the garage slab to help prevent cracking. Fig. 4-14 shows the base material prepared for the placement of concrete.

Stairway requirements include minimum sizes of tread and maximum sizes of risers. The tread is the part of the stairs that the foot is placed on; the riser is the vertical piece of wood between the treads. The applicable codes should be checked to determine the acceptable sizes of the thread and risers. Fig. 4-15 shows the requirements for dimensions for a stairway for most codes.

Figure 4-14 Base material for concrete slab

Figure 4-15 Stair dimensions

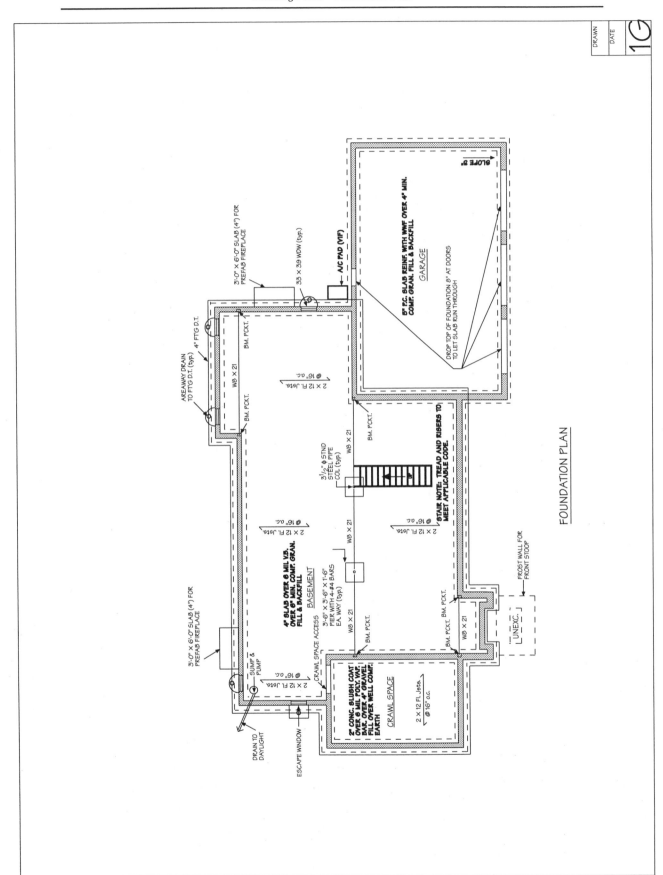

FOUNDATION PLAN

GARAGE

8" P.C. SLAB REINF. WITH WWF OVER 4" MIN.
COMP. GRAN. FILL & BACKFILL

SLOPE B"

DROP TOP OF FOUNDATION 8" AT DOORS
TO LET SLAB RUN THROUGH

A/C PAD (VIF)

33 × 39 WDW (typ.)

3'-0" × 6'-0" SLAB (4") FOR
PREFAB FIREPLACE

4" FTG D.T.

AREAWAY DRAIN
TO FTG D.T. (typ.)

BM. PCKT.

W8 × 21

BM. PCKT.

2 × 12 FL Jsts.
@ 16" o.c.

W8 × 21

BM. PCKT.

3½" ⌀ STND
STEEL PIPE
COL (typ.)

STAIR NOTE: TREAD AND RISERS TO
MEET APPLICABLE CODE.

W8 × 21

2 × 12 FL Jsts.
@ 16" o.c.

FROST WALL FOR
FRONT STOOP

UNEXC.

BM. PCKT.

W8 × 21

BM. PCKT. BM. PCKT.

BM. PCKT.

CRAWL SPACE

2 × 12 FL Jsts.
@ 16" o.c.

2" CONC. SLUSH COAT
OVER 6 MIL POLY. VAP.
BAR. OVER 4" GRAVEL
FILL OVER WELL COMP.
EARTH

CRAWL SPACE ACCESS

BASEMENT

4" SLAB OVER 6 MIL V.B.
OVER 6" MIN. COMP. GRAN.
FILL & BACKFILL

2 × 12 FL Jsts.
@ 16" o.c.

3'-6" × 3'-6" × 1'-6"
PIER WITH 4-#4 BARS
EA. WAY (typ.)

2 × 12 FL Jsts.
@ 16" o.c.

SUMP &
PUMP

3'-0" × 6'-0" SLAB (4") FOR
PREFAB FIREPLACE

DRAIN TO
DAYLIGHT

ESCAPE WINDOW

DRAWN

DATE

1G

Drawing 1H—Utilities and Section Marks

The majority of the utilities are located in the basement of the house. The utilities included in this design drawing are heat, water, wastewater treatment, hot water, and electric service.

Since the house has over 3000 square feet of heated floor area, two independent furnaces are specified. The first floor of the house is living units and the second floor is sleeping units. Each of the furnaces services these two areas independently.

The water supply for this house is a well system. A shaft is drilled until a water source that is sufficient to meet anticipated demand is located. The water is pumped from the well shaft to a water softening unit in the basement for conditioning.

A water heater is placed adjacent to the two furnaces to keep the utilities in one location and to utilize one exhaust ventilation stack. The capacity required for the water heater is a function of the number of bathrooms.

The wastewater from the house must be either disposed of on site (septic system) or sent for treatment to a wastewater plant. If the latter is chosen, it is simply a matter of connecting the main drain pipe to the municipality trunk line. A septic system (Fig. 4-16) however, requires considerably more site work. A septic system was used for this design. Design of the septic system traditionally is not included in the design drawings for construction of the house.

Section marks A, B, C, and D are shown on this drawing. Detailed sections at these locations are presented on page 4 of the set of drawings and are discussed later in this publication.

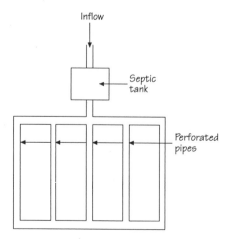

Figure 4-16 Schematic of septic sewer system

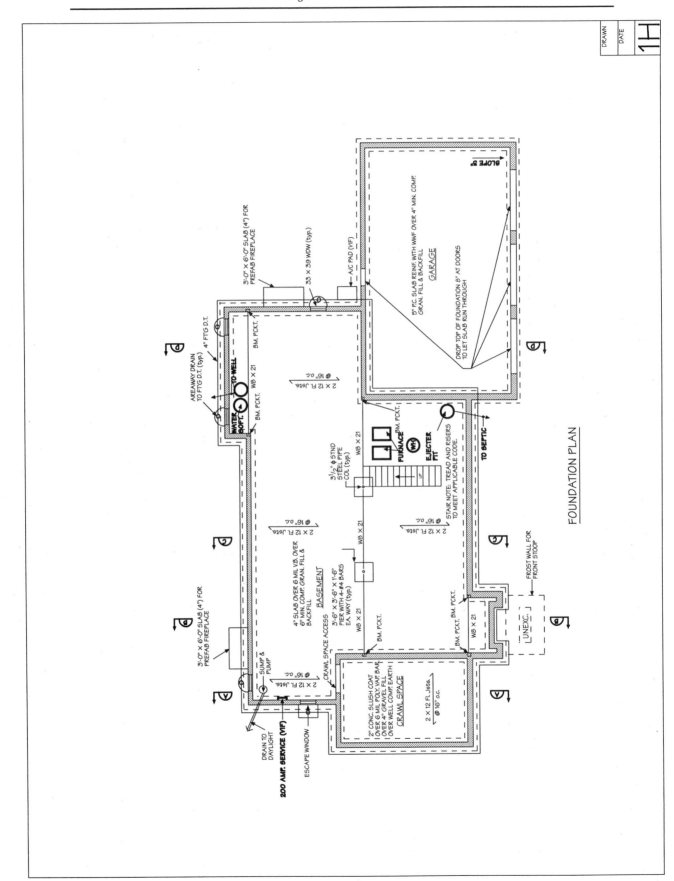

FOUNDATION PLAN

Drawing 1I—General Notes

General notes provide clarification and prevent miscommunication among designer, owner, and contractor. Note that every bit of information need not be listed on the drawings, since there will also be a contract and specifications to help the parties understand their respective duties. The general notes on this drawing serve to let the contractor know the basis of the engineer's design (allowable soil load, header sizes), information about the material (type of wood to be used, minimum concrete strength requirements), and notifying the contractor that regardless of what the drawings show, the building codes must be satisfied. Other information can be placed in the general notes for example:

- Steel beam and pipe column specifications.

- Stipulation that drawn dimensions take precedence over scaling.

- Requirements to be met for material to be supplied by plumbers, electricians, and heating subcontractors.

- Disclaimer by architect that contractor must verify dimensions.

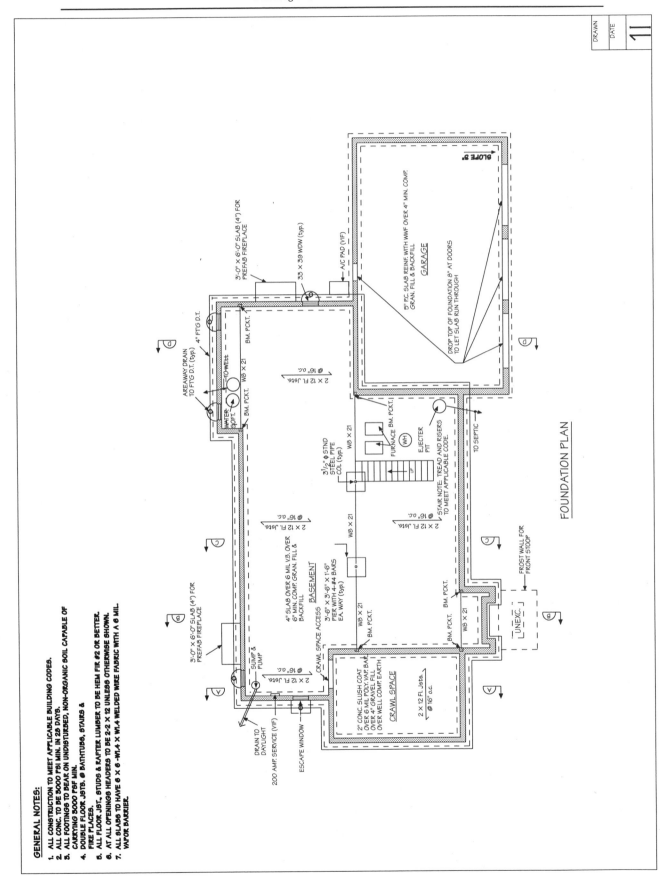

FOUNDATION PLAN

GENERAL NOTES:

1. ALL CONSTRUCTION TO MEET APPLICABLE BUILDING CODES.
2. ALL CONC. TO BE 3000 PSI MIN. IN 28 DAYS.
3. ALL FOOTINGS TO BEAR ON UNDISTURBED, NON-ORGANIC SOIL CAPABLE OF CARRYING 3000 PSF MIN.
4. DOUBLE FLOOR JSTS. @ BATHTUBS, STAIRS & FIRE PLACES.
5. ALL FLOOR JST., STUDS & RAFTER LUMBER TO BE HEM FIR #2 OR BETTER.
6. AT ALL OPENINGS HEADERS TO BE 2-2 × 12 UNLESS OTHERWISE SHOWN.
7. ALL SLABS TO HAVE 6 × 6 -W1.4 × W1.4 WELDED WIRE FABRIC WITH A 6 MIL VAPOR BARRIER.

Drawing 1J—Electrical

The electrical system for the basement is usually not very elaborate and only includes pull chain lights, smoke detectors, and electrical outlets. The symbols show the many different components of the electrical system that might be used are listed below.

	Light switch
	110 V duplex outlet
	110V duplex outlet with ground fault
	Exhaust fan
	Timer
	Smoke detector
	Porcelain lamp holder with pull chain
	Surface mounted incand. fixture
	Recessed lighting

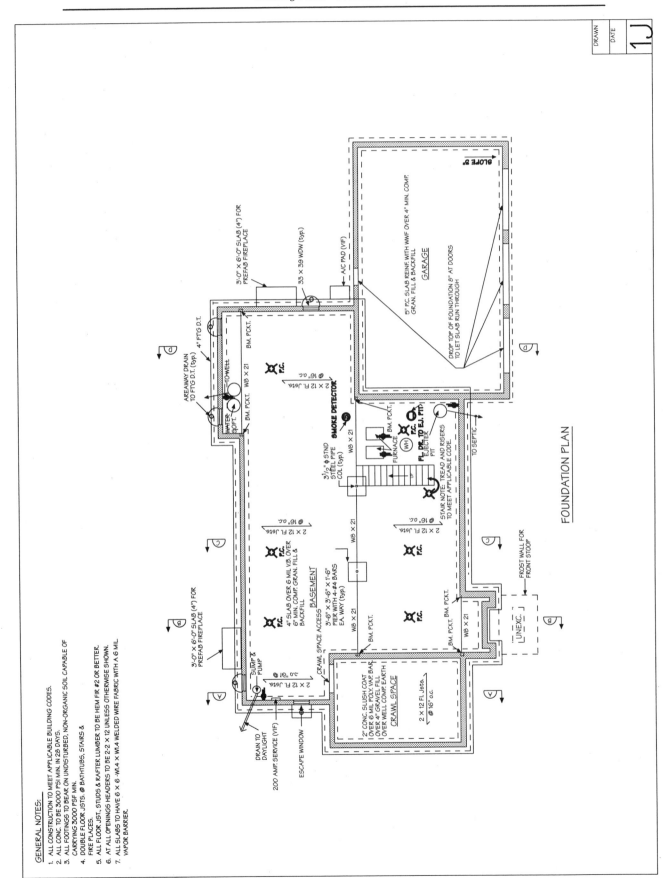

FOUNDATION PLAN

GENERAL NOTES:

1. ALL CONSTRUCTION TO MEET APPLICABLE BUILDING CODES.
2. ALL CONC. TO BE 3000 PSI MIN. IN 28 DAYS.
3. ALL FOOTINGS TO BEAR ON UNDISTURBED, NON-ORGANIC SOIL CAPABLE OF CARRYING 3000 PSF MIN.
4. DOUBLE FLOOR JSTS. @ BATHTUBS, STAIRS & FIRE PLACES.
5. ALL FLOOR JST., STUDS & RAFTER LUMBER TO BE HEM FIR #2 OR BETTER.
6. AT ALL OPENINGS HEADERS TO BE 2-2 × 12 UNLESS OTHERWISE SHOWN.
7. ALL SLABS TO HAVE 6 × 6 -W1.4 × W1.4 WELDED WIRE FABRIC WITH A 6 MIL VAPOR BARRIER.

Drawings 1K and 1L—Exterior and Interior Dimensions

Dimensioning is very straightforward and needs no discussion. It should be noted that exterior wall dimensions are taken at the outside of the face brick or concrete. Inside dimensions are taken at the face of the wood studs.

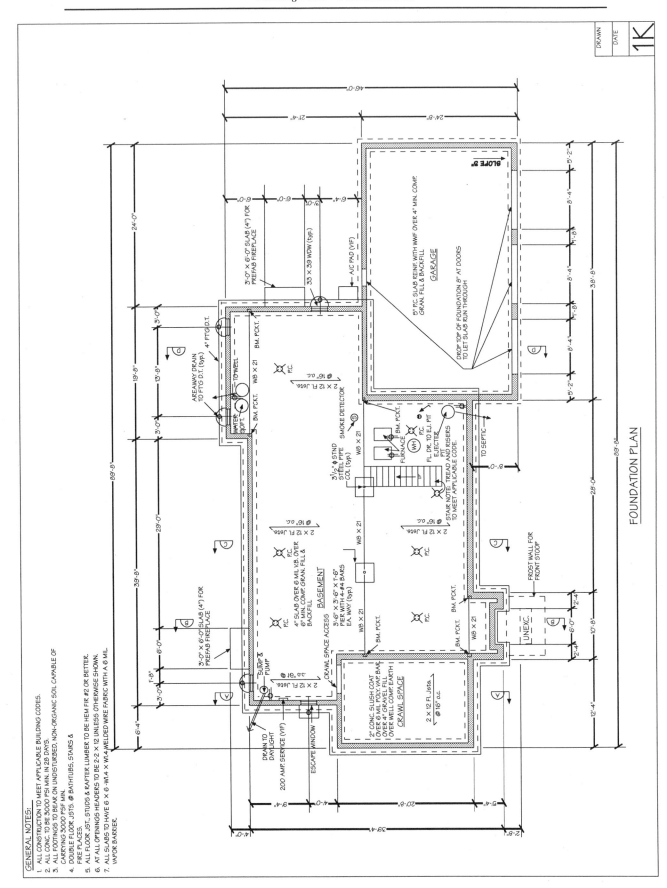

GENERAL NOTES:
1. ALL CONSTRUCTION TO MEET APPLICABLE BUILDING CODES.
2. ALL CONC. TO BE 3000 PSI MIN. IN 28 DAYS.
3. ALL FOOTINGS TO BEAR ON UNDISTURBED, NON-ORGANIC SOIL CAPABLE OF CARRYING 3000 PSF MIN.
4. DOUBLE FLOOR JSTS. @ BATHTUBS, STAIRS & FIRE PLACES.
5. ALL FLOOR JST, STUDS & RAFTER LUMBER TO BE HEM FIR #2 OR BETTER.
6. AT ALL OPENINGS HEADERS TO BE 2-2 x 12 UNLESS OTHERWISE SHOWN.
7. ALL SLABS TO HAVE 6 x 6 -W1.4 x W1.4 WELDED WIRE FABRIC WITH A 6 MIL VAPOR BARRIER.

FOUNDATION PLAN

1K

DRAWN
DATE

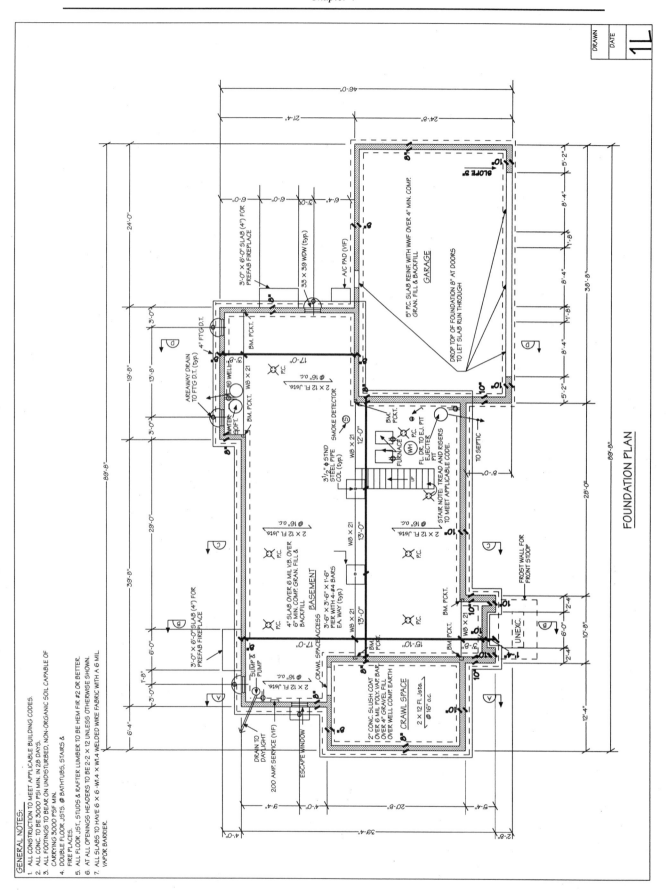

FOUNDATION PLAN

GENERAL NOTES:
1. ALL CONSTRUCTION TO MEET APPLICABLE BUILDING CODES.
2. ALL CONC. TO BE 3000 PSI MIN. IN 28 DAYS.
3. ALL FOOTINGS TO BEAR ON UNDISTURBED, NON-ORGANIC SOIL CAPABLE OF CARRYING 3000 PSF MIN.
4. DOUBLE FLOOR JSTS @ BATHTUBS, STAIRS & FIRE PLACES.
5. ALL FLOOR JST., STUDS & RAFTER LUMBER TO BE HEM FIR #2 OR BETTER.
6. AT ALL OPENINGS HEADERS TO BE 2-2 x 12 UNLESS OTHERWISE SHOWN.
7. ALL SLABS TO HAVE 6 x 6 -W1.4 x W1.4 WELDED WIRE FABRIC WITH A 6 MIL VAPOR BARRIER.

2—First Floor Plan

Drawing 2A—Exterior Wood Framing

Drawing 2B—Interior Wood Framing

Drawing 2C—Doors and Stairs

Drawing 2D—Kitchen Layout

Drawing 2E—Utility Room Layout

Drawing 2F—Brick Veneer and Prefabricated Fireplace

Drawing 2G—Windows

Drawing 2H—Electrical Outlets

Drawing 2I—Fixtures

Drawing 2J—Electrical Wiring

Drawing 2K—Ceiling Joists and Pre-engineered Trusses

Drawing 2L—Utilities and Header Beams

Drawing 2M—Exterior Dimensions

Drawing 2N—Interior Dimensions

Drawing 2A and 2B—Exterior and Interior Wood Framing

Drawings 2A and 2B show the locations of the wood frame walls. The walls can be exterior, bearing, and partition walls. Exterior walls support the roof and its accompanying loads, as well as a portion of the floor and its loads. The bearing walls utilized in this structure carry only the load of walls and floors and typically do not support any of the roof loads. Partition walls carry only their own weight and are identified by the absence of any joists supported on the wall. Bearing and exterior walls are best supported directly on the basement steel beams and columns in the basement to avoid deflection problems.

The majority of walls are constructed of 2 × 4s. In some cases where extra wall width or strength is needed, 2 × 6s may be used instead. Walls should be constructed using spacing and sizes of members that are industry standards. Deviation from the norm may result in problems of attaching cladding, plaster board, and trim. The standard spacing of the vertical members (studs) in walls is 16 inches.

There are two types of framing, platform and balloon, as shown in Figs. 4-17 and 4-18. Note that the difference between balloon and platform framing is that the studs in the former are continuous up to the second floor; the latter has a break in the studs at the bottom of the second floor. Platform framing is the most common system and is used for this design. The frames can be built on site and then tilted up into place. Or the frames can be built at a plant and shipped in trucks to the site. The frames are lifted from the truck into place by the use of a crane.

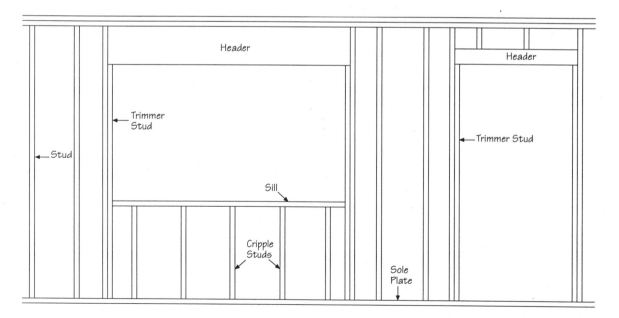

Figure 4-17 Platform framing

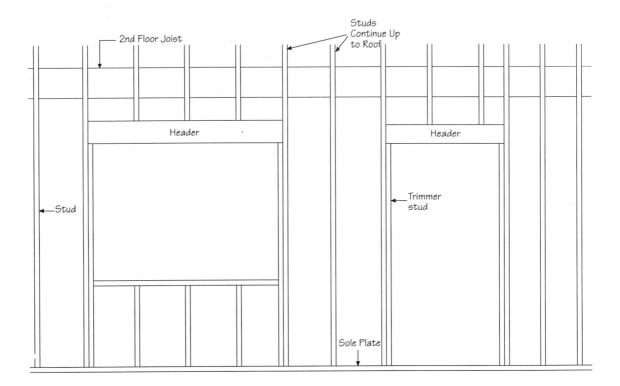

Figure 4-18 Balloon framing

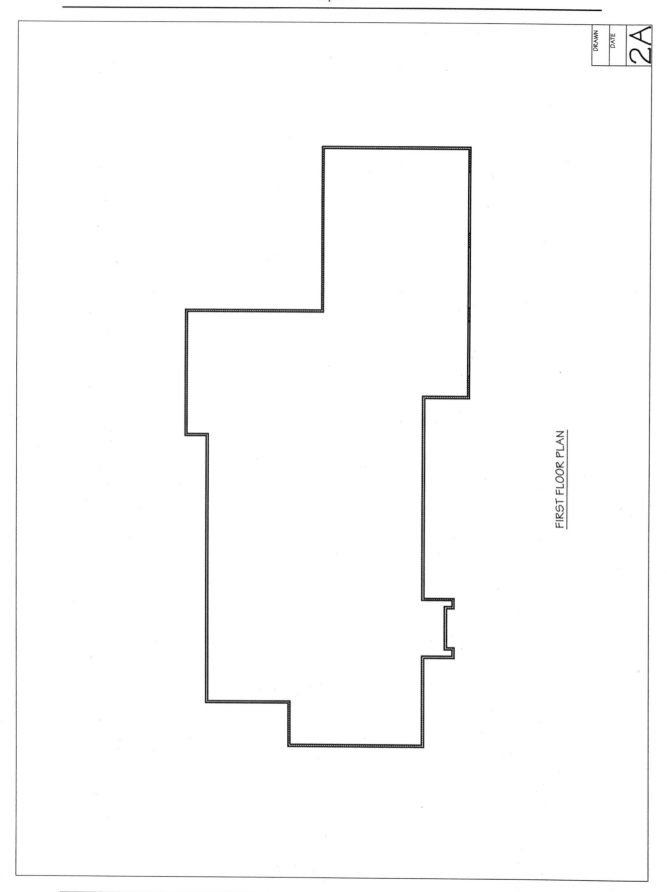

FIRST FLOOR PLAN

DRAWN

DATE

2A

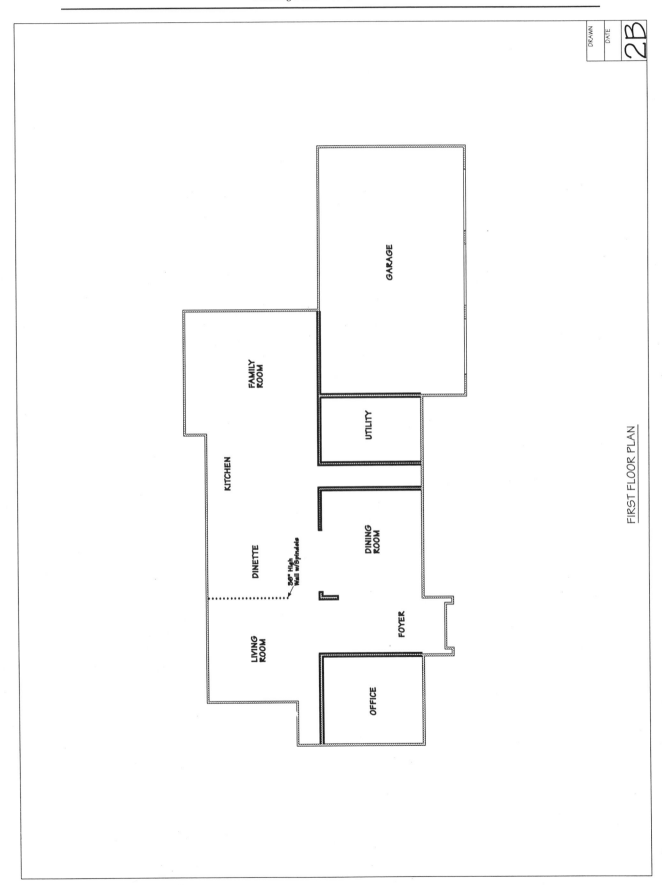

FIRST FLOOR PLAN

Drawing 2C, 2D, and 2E—Doors, Stairs, Kitchen Layout, and Utility Layout

The details provided on these drawings are self-evident and need little explanation. The appliances require a manufacturer specified amount of clearances. Sample clearances for a stove are shown in Fig. 4-19.

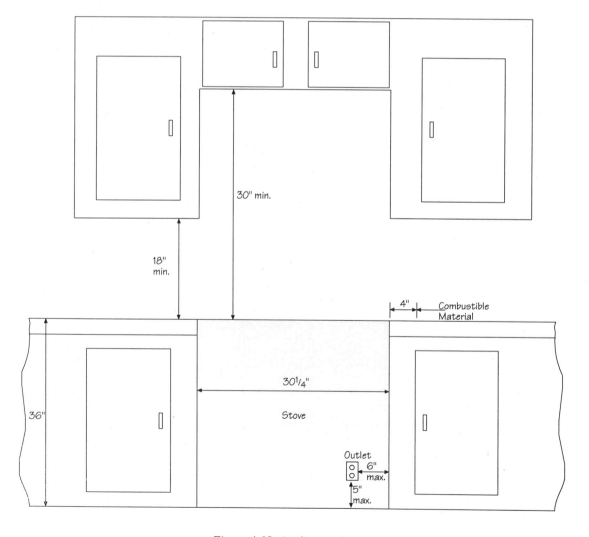

Figure 4-19 Appliance clearance

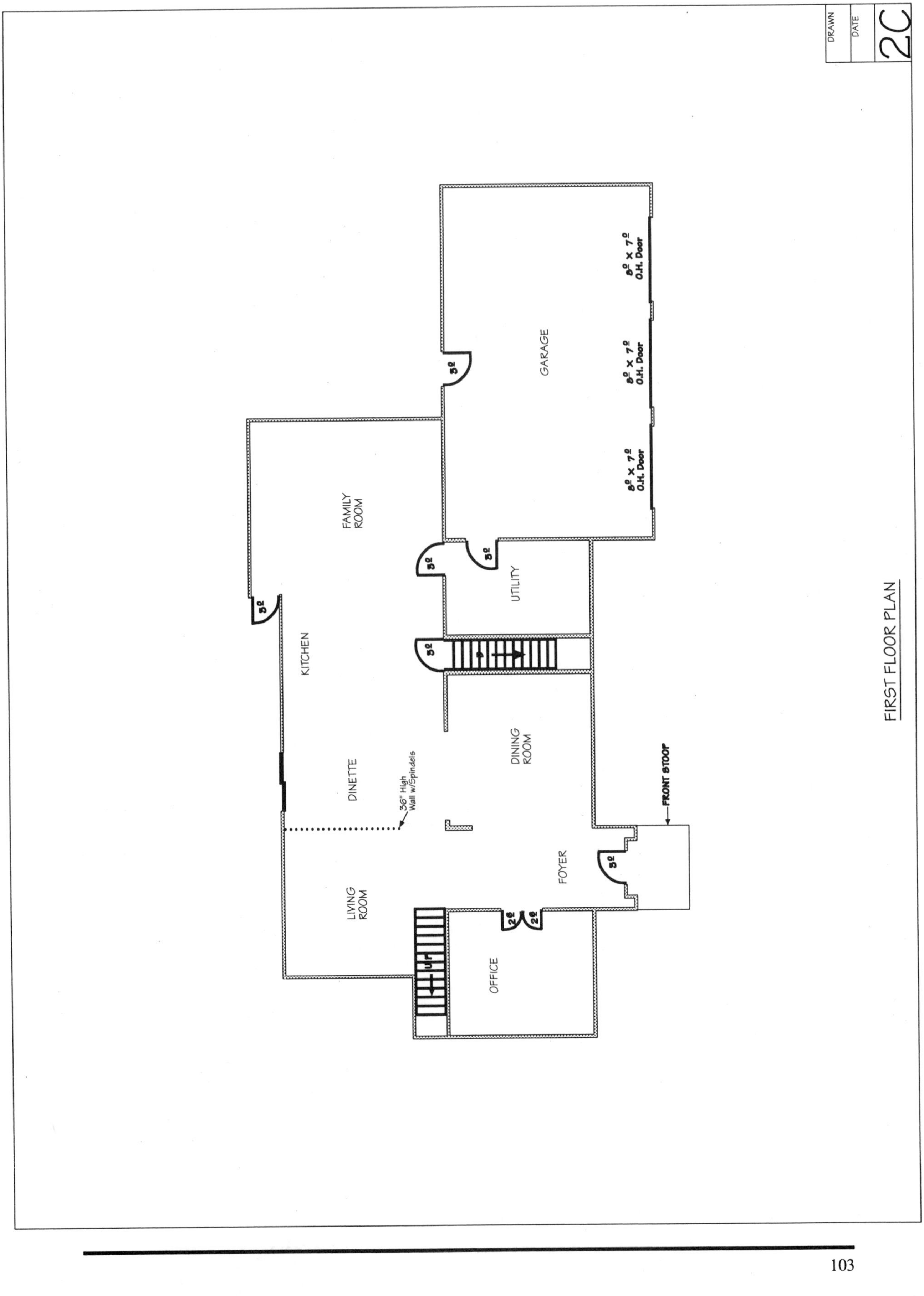

FIRST FLOOR PLAN

GARAGE

$8^0 \times 7^0$
O.H. Door

$8^0 \times 7^0$
O.H. Door

$8^0 \times 7^0$
O.H. Door

FAMILY
ROOM

3^2

3^2

KITCHEN

3^2

UTILITY

3^2

3^2

DINETTE

3'6" High
Wall w/Spindels

DINING
ROOM

LIVING
ROOM

UP

DN

FOYER

3^2

2^6

2^6

OFFICE

FRONT STOOP

DRAWN

DATE

2C

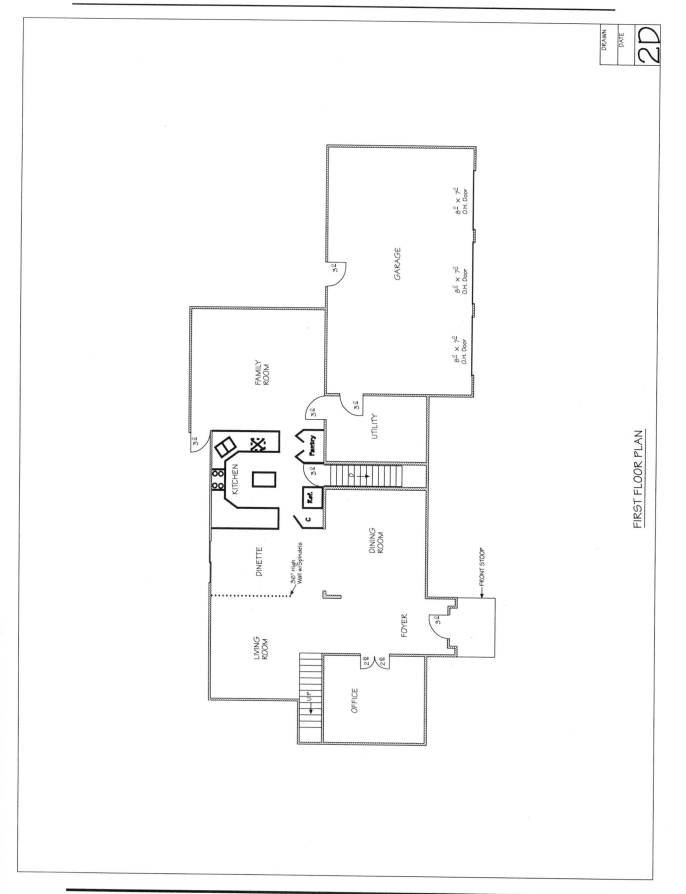

FIRST FLOOR PLAN

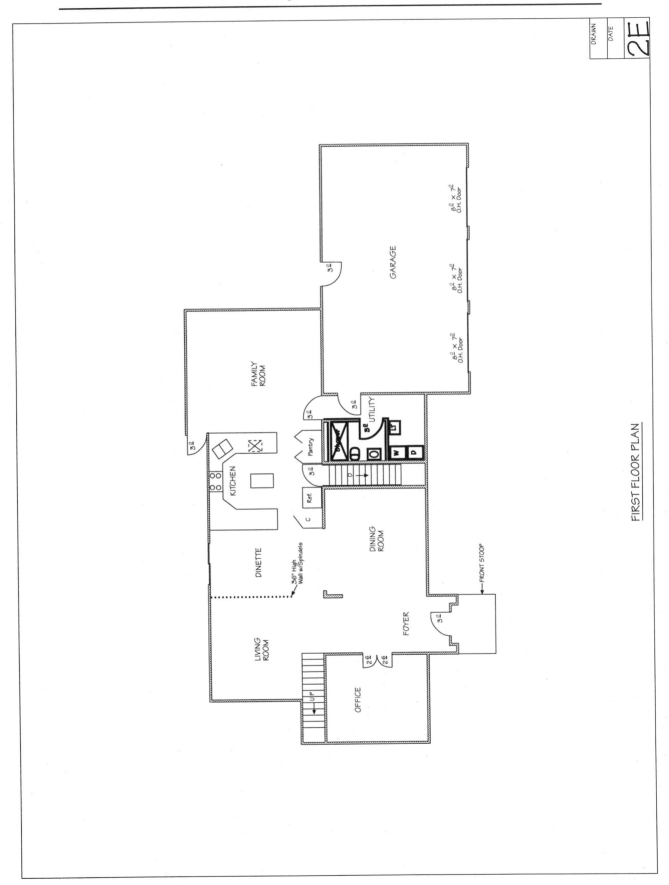

FIRST FLOOR PLAN

Drawing 2F—Brick Veneer and Prefabricated Fireplace

Brick veneer is used for the front elevation of this structure. The foundation supporting the brick veneer must be wider than if wood or aluminum siding is used so that the brick can be supported directly on the foundation. The plan view of the structure as shown here specifies where the brick goes, but does not specify the mortar joint type, or the style for laying of the brick. These details may be covered in the specifications or on elevation drawings. The three most popular mortar joints, as well as the two styles for laying bricks, are shown below in Figs. 4-20 and 4-21.

A prefabricated steel fireplace is used in this design. The unit is secured by nailing into an opening of the framing and placing it on top of a concrete pad.

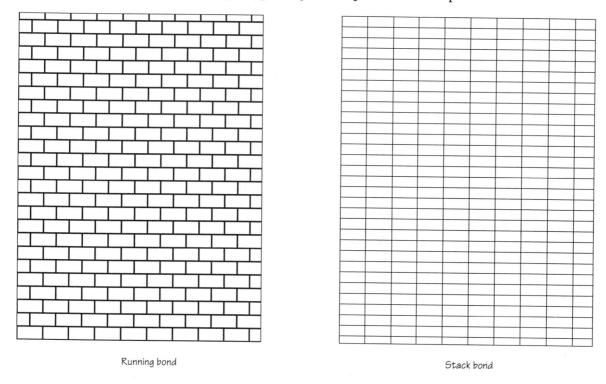

Running bond   Stack bond

Figure 4-20 Brick patterns

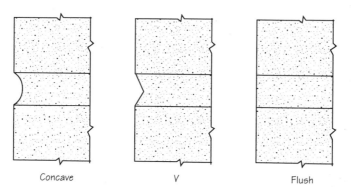

Concave   V   Flush

Figure 4-21 Mortar joints

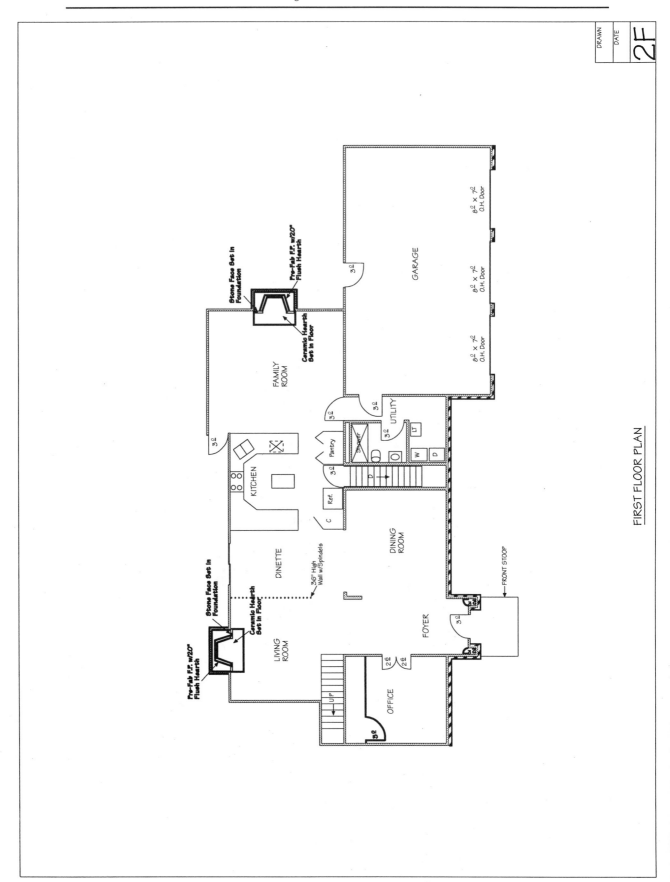

FIRST FLOOR PLAN

Drawing 2G—Windows

This drawing shows the location of the windows on the first floor. The type and size of window to be installed is not determined from this plan view, but from the elevation (see drawings 5 and 6). This drawing is only used to determine where the windows will be located. However, types and styles of windows are discussed here.

There are many types of windows and many manufacturers. Two of the most popular types of windows are the double hung and casement windows. The bottom sash of the double hung window slides up for ventilation. In contrast, ventilation is obtained by rotation of one side of the casement window. A casement window is shown in Fig. 4-22.

The window size required is determined by a combination of the owner's requirements and satisfaction of the applicable building code. Figure 4-23 shows how window model numbers are designated and how the sash opening, glass size, and required rough opening are determined from that model number.

Figure 4-22 Casement windows

A window designated as 3846 denotes the following:
sash width opening = 3'-8"
sash height opening = 4'-6"
glass width size = sash width -3½" = 3'-4½"
glass height size = (sash height -6")/2 = 2'-0"
rough width opening = sash opening + 2" = 3'-10"
rough height opening = sash opening + 3" = 4'-0"

Figure 4-23 Window opening calculations

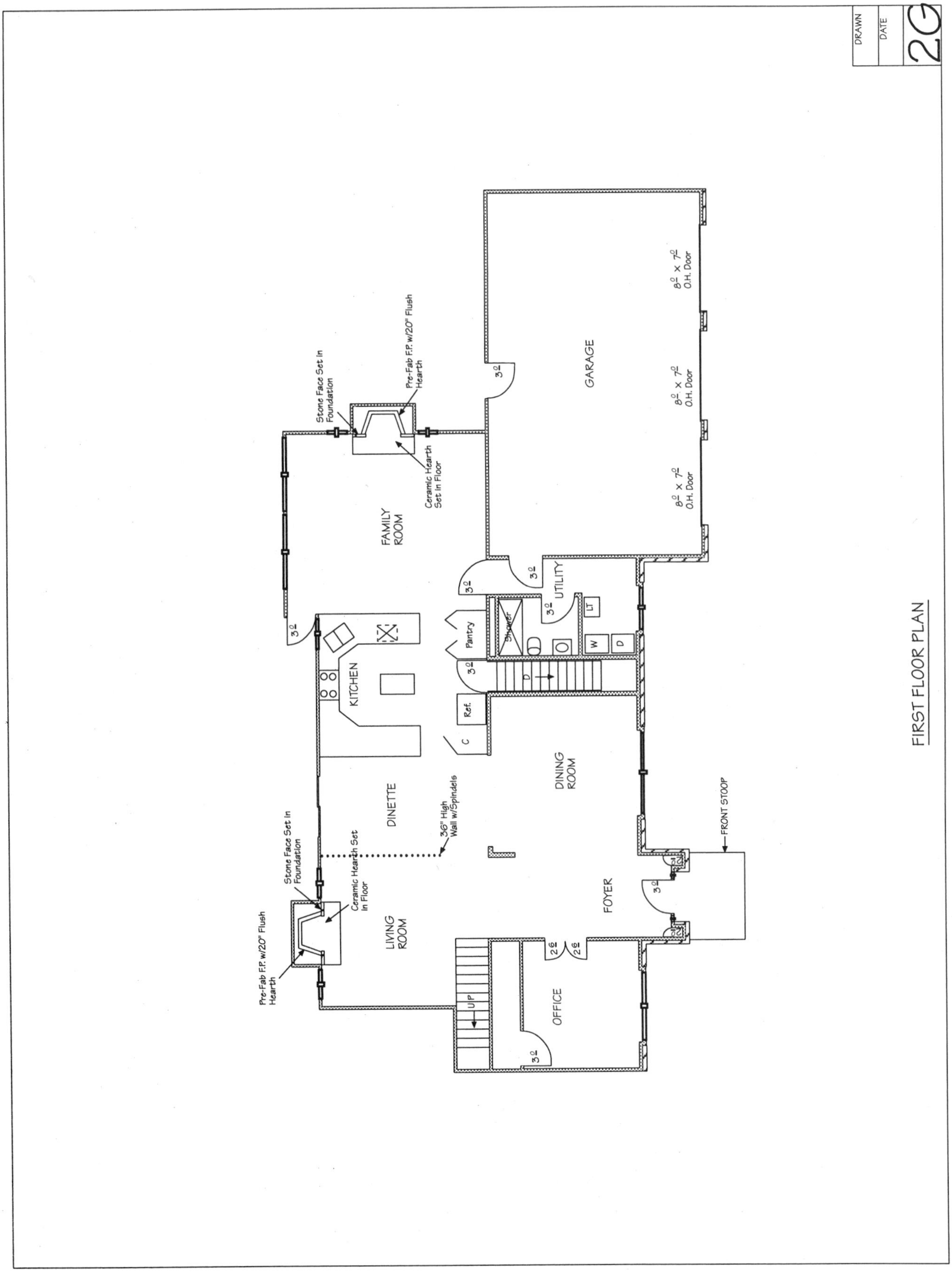

FIRST FLOOR PLAN

DRAWN
DATE
2G

GARAGE

$8^{0} \times 7^{0}$ O.H. Door

$8^{0} \times 7^{0}$ O.H. Door

$8^{0} \times 7^{0}$ O.H. Door

FAMILY ROOM

Stone Face Set In Foundation

Pre-Fab F.P. w/20" Flush Hearth

Ceramic Hearth Set In Floor

KITCHEN

DINETTE

36" High Wall w/Spindels

Stone Face Set In Foundation

Pre-Fab F.P. w/20" Flush Hearth

Ceramic Hearth Set In Floor

LIVING ROOM

UTILITY

LT

W

D

Pantry

C

Ref.

C

DINING ROOM

FOYER

FRONT STOOP

UP

OFFICE

2^{6}

2^{6}

Drawings 2H, 2I, and 2J—Electrical Outlets, Fixtures, and Electrical Wiring

The locations of electrical outlets are shown on drawing 2H. Ground fault indicators (GFI) outlets must be used where water or moisture may be present. GFI outlets have the capability of cutting off power to the outlet if moisture is detected. If the outlet is for exterior use, it must be waterproof (WP).

The spacing of electrical outlets is controlled by the local building code. Outlets must not be farther than 12 feet from another outlet. This eliminates the hazard of long extension cords that are tripping as well as fire hazards.

Drawings 2I and 2J show the locations of the light fixtures and the switches. The more complicated the wiring, the more expensive it will be. The extent of the wiring needed is based on the requirements of the home owner. Note that the lines shown between switches and outlets do not represent the actual location of electric lines. These lines are only for the purpose of showing which outlets and switches are interconnected.

FIRST FLOOR PLAN

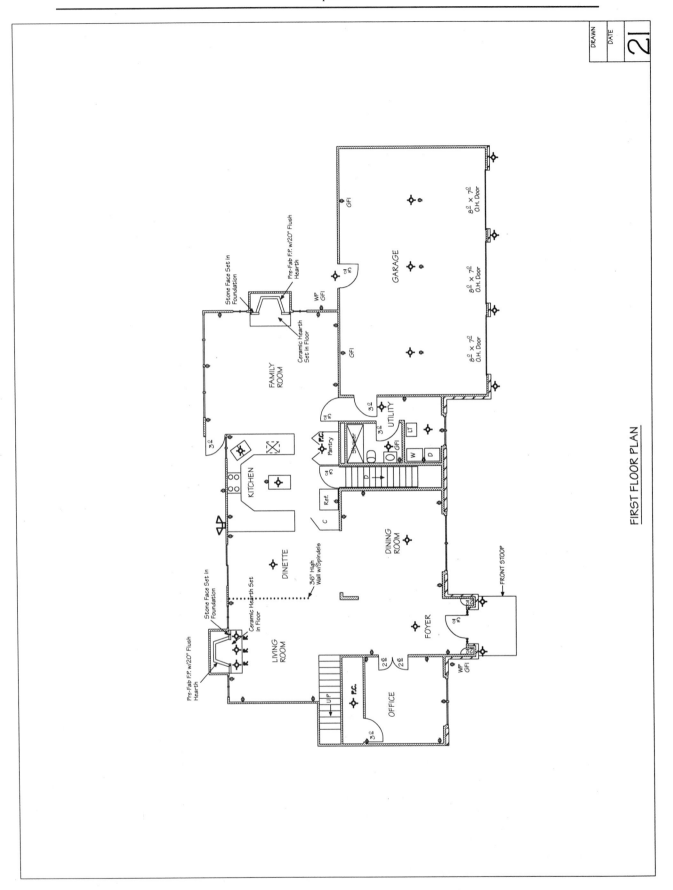

FIRST FLOOR PLAN

FIRST FLOOR PLAN

Drawing 2K—Ceiling Joists and Pre-Engineered Trusses

The ceiling joists shown on this drawing are actually the floor joists for the second floor. Floor joists were discussed previously in Drawing 1E and the reader should refer to the corresponding text for further information. Note that the long span length over the family room (only the portion below the second story) precluded the use of 2 × 12s at 16" on center. When this situation occurs, the depth of joists can be increased or the spacing can be decreased. For this design, the spacing was decreased since an increase in the joist depth would result in a different ceiling level in a portion of the family room than the rest of the structure.

A portion of the first floor does not have a second level above it, but is directly below the roof. The roof system is designated on the drawing as "Pre-engineered Trusses." The truss appears, and is often treated as a simple system to analyze. However, truss behavior is very complicated and analysis is usually performed by the truss supplier, with sophisticated truss design software. Pre-engineered trusses are factory assembled, which allows a precisely built structural member. The trusses are shipped to the site and lifted into place with a crane.

Three typical trusses are shown in Fig. 4-24. The Pratt and Howe trusses can be used for houses with typical flat ceilings, whereas, the scissors truss can be used for houses with vaulted ceilings.

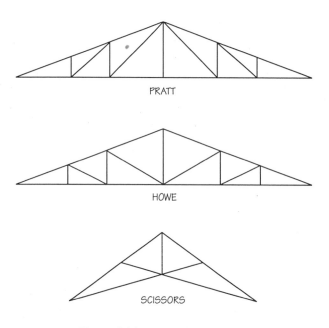

Figure 4-24 Types of roof trusses

FIRST FLOOR PLAN

2K
DRAWN
DATE

GARAGE

Pre-Engineered
Wood Trusses at 24"

8⁰ × 7⁰
O.H. Door

8⁰ × 7⁰
O.H. Door

8⁰ × 7⁰
O.H. Door

8⁰ × 7⁰
O.H. Door

GFI

Pre-Engineered Wood Trusses at 24"

GFI

2 × 12 Clg Joists
at 16" o.o.

UTILITY

W
D

LT

GFI

P.C.

Stone Face Set In
Foundation

Pre-Fab F.P. w/20" Flush
Hearth

Ceramic Hearth
Set In Floor

FAMILY
ROOM

Double 2 × 12 Clg
Joists at 12" o.o.

WP
GFI

Pantry

2 × 12 Clg Joists
at 16" o.o.

KITCHEN

C

Ref.

DINING
ROOM

2 × 12 Clg Joists
at 16" o.o.

DINETTE

36" High
Wall w/Spindels

2 × 12 Clg Joists
at 16" o.o.

Stone Face Set In
Foundation

Pre-Fab F.P. w/20" Flush
Hearth

Ceramic Hearth
Set In Floor

Vaulted
Ceiling

LIVING
ROOM

UP

P.C.

OFFICE

2 × 12 Clg Joists
at 16" o.c.

Vaulted
Ceiling

FOYER

FRONT STOOP

WP
GFI

2⁶

2⁶

115

Drawing 2L—Utilities and Header Beams

The utilities shown on this drawing include gas lines provided with shutoffs, freeze-proof hose bibs (F.P.H.B) for external water supply, and a bathroom exhaust fan.

A gas curb is located between the garage and the living space. The gas curb is actually a concrete step at the threshold of the house. Automobile exhaust in the garage tends to accumulate at ground level. The difference in elevation provided by the gas curb prevents the exhaust from flowing freely into the house.

Openings in walls for doors, windows, and passageways can be a problem because they interrupt the structural load carrying system. Beams must be provided over these openings so loads above the them are transferred across the opening to the adjacent studs. These beams are called headers. Two types of header beams are shown on the drawing, a 2 × 12 with ½" plywood header and a 1¾ × 11⅞ laminated beam. A crossection of the 2 × 12 with ½" plywood is shown in Fig. 4-25. The crossection of the laminated beams is shown in Fig. 4-26. A laminated member is manufactured by taking thin pieces of wood and gluing them together with the grain in a predetermined pattern. A heat and pressure treatment is used to permanently bond the material. This provides a beam with uniform properties and with a strength of three times that of a conventional beam.

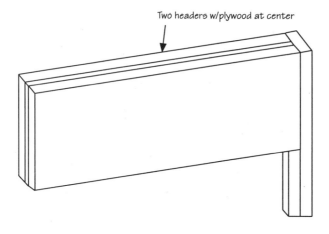

Two headers w/plywood at center

Figure 4-25 2 × 12 with ½" plywood header

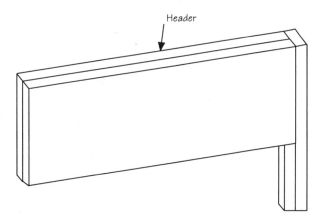

Figure 4-26 Laminated header

FIRST FLOOR PLAN

Drawings 2M and 2N—Exterior and Interior Dimensions

Refer to drawings 1K and 1L.

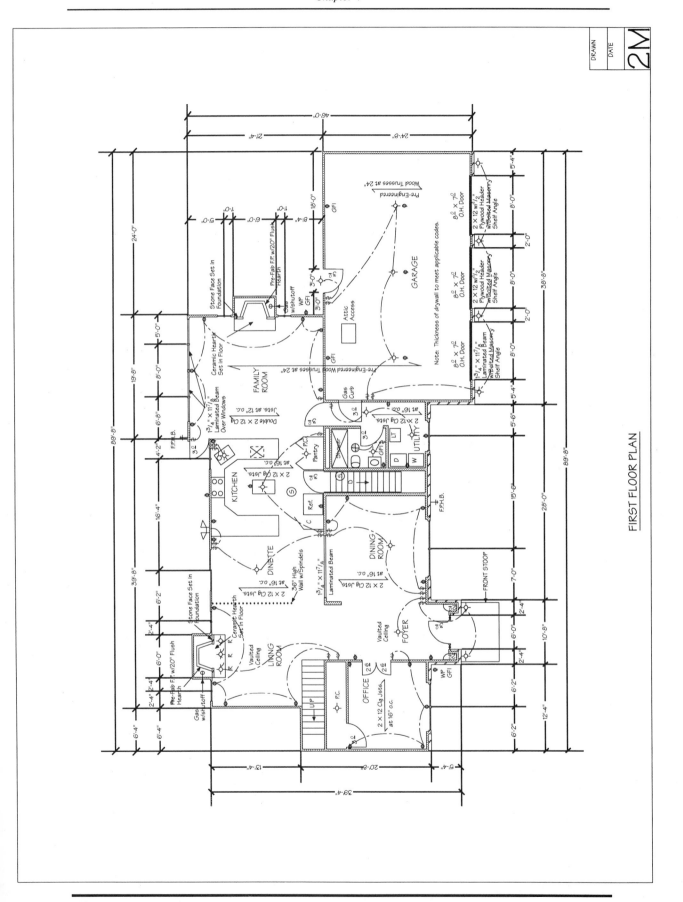

FIRST FLOOR PLAN

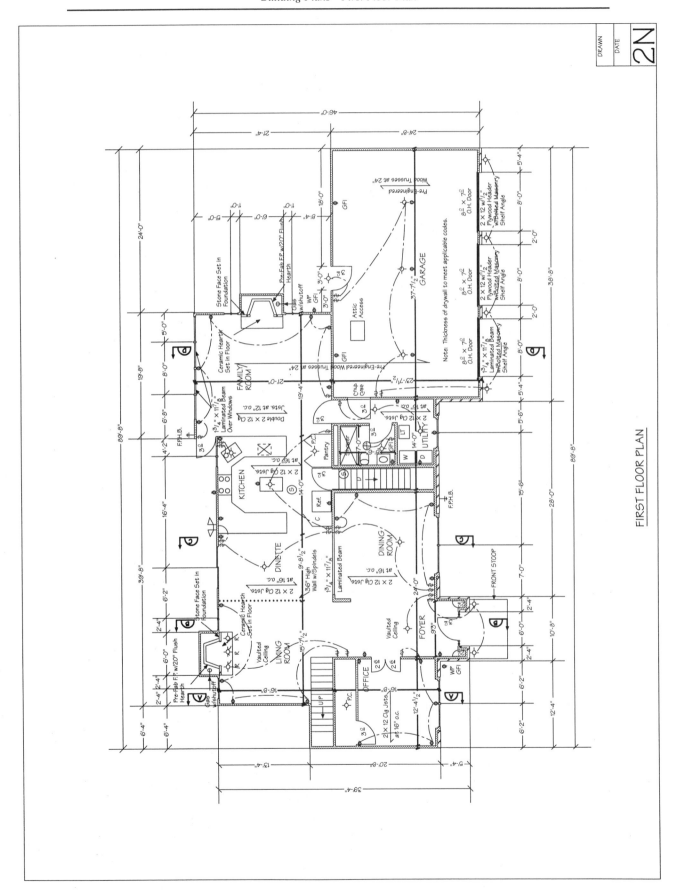

FIRST FLOOR PLAN

3—Second Floor Plan

Drawing 3A—Layout

Drawing 3B—Doors, Railings, and Stairs

Drawing 3C—Closets

Drawing 3D—Brick Face and Bathroom Layout

Drawing 3E—Windows, Roofing, Trusses, and Laminated Beams

Drawing 3F—Electrical Plan

Drawing 3G—Interior Dimensions

Drawing 3H—Exterior Dimensions

The drawings that compose the second floor plan are very similar to the drawings for the first floor plan. For the sake of completeness, the chronology of the second floor plan drawings is provided. For information on the individual drawings for the second floor plan, consult the references and discussion presented below.

Drawing 3A—Layout

Refer to drawings 2A and 2B.

Drawing 3B—Doors, Railings, and Stairs

This drawing shows the width of door openings as well as the direction the doors open.

Drawing 3C—Closets

Closet space is dictated by the needs of the owner. Closets are placed so they will not disrupt the shape of the room.

Drawing 3D—Brick Face and Bathroom Layout

Refer to drawings 2E and 2F.

Placing pipes in external walls should be avoided because it represents a possible freezing hazard. Bathroom plumbing fixtures should be placed within the bathroom to avoid this situation.

Drawing 3E—Windows, Roofing, Trusses, and Laminated Beams

Refer to drawings 2G, 2K, and 2L.

The horizontal lines on the right portion of the structure represent the roof over the first story.

Drawing 3F—Electrical Plan

Refer to drawing 2J.

Drawing 3G—Interior Dimensions

Refer to drawing 2N.

Drawing 3H—Exterior Dimensions

Refer to drawing 2M.

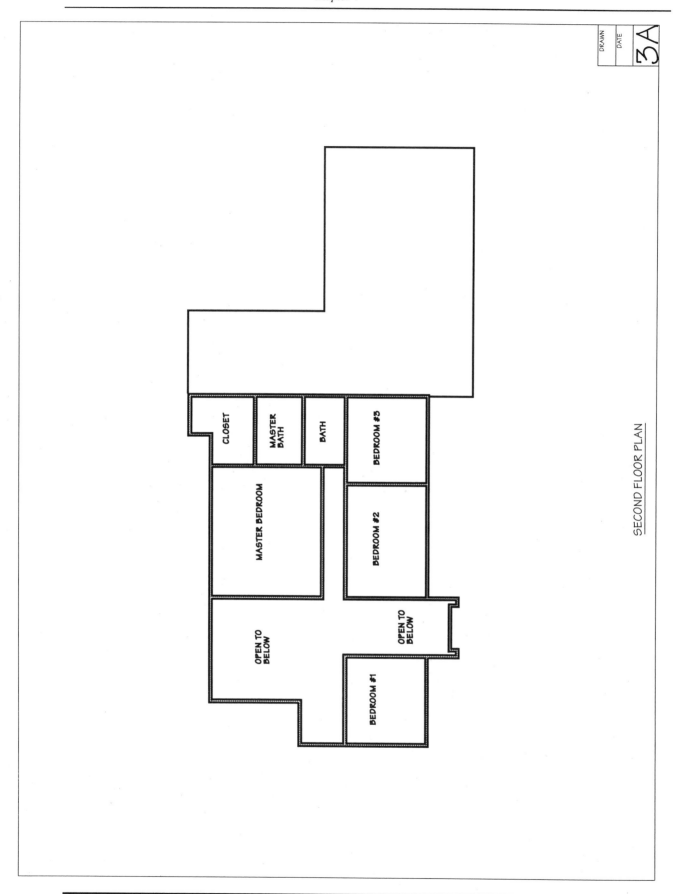

SECOND FLOOR PLAN

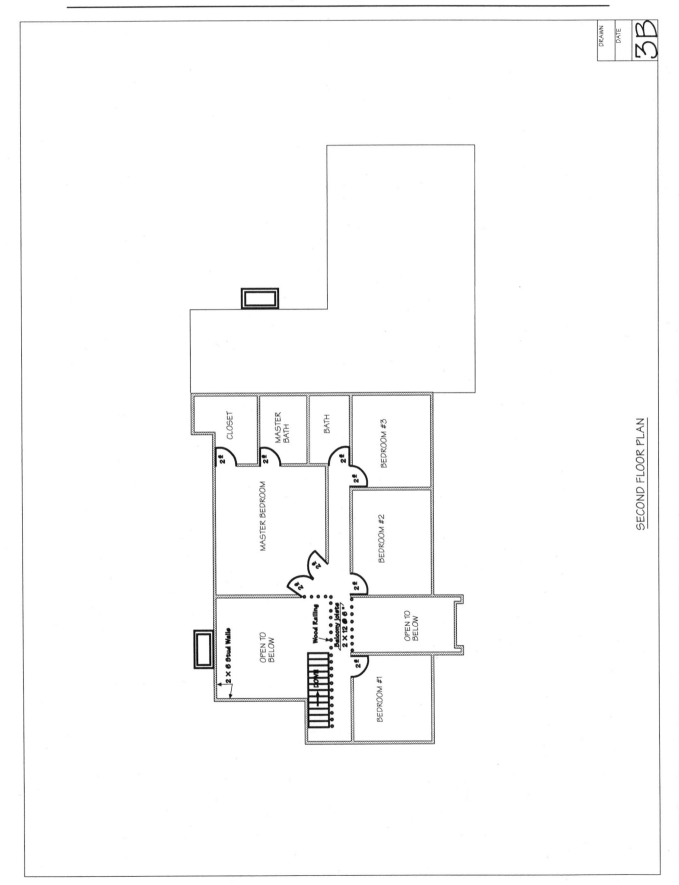

SECOND FLOOR PLAN

3B

DRAWN

DATE

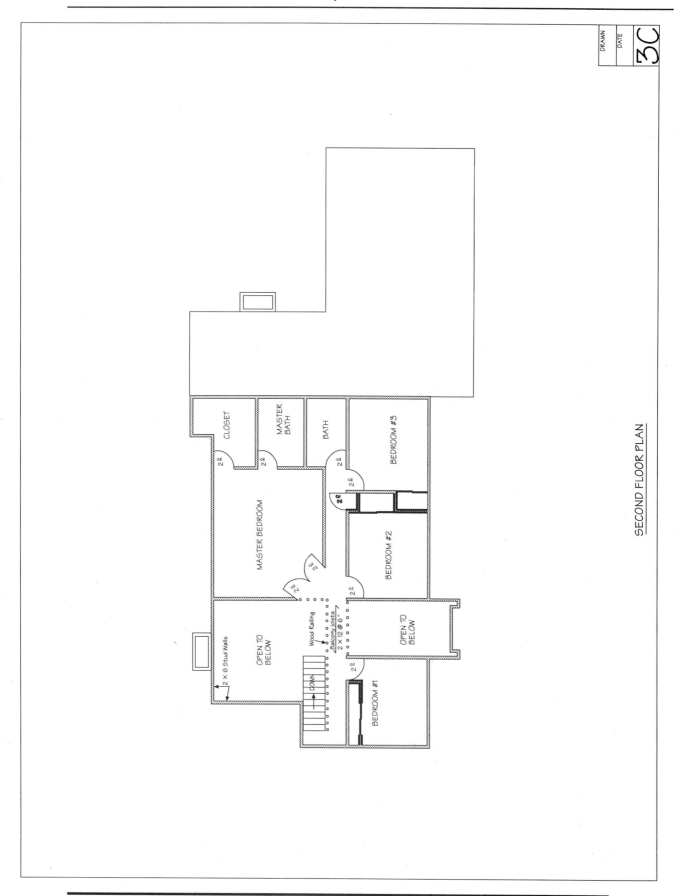

SECOND FLOOR PLAN

CLOSET

MASTER BATH

BATH

BEDROOM #3

MASTER BEDROOM

BEDROOM #2

OPEN TO BELOW

2 × 6 Stud Walls

Wood Railing

Balcony Joists
2 × 12 @ 6"

OPEN TO BELOW

DOWN

BEDROOM #1

DRAWN

DATE

3C

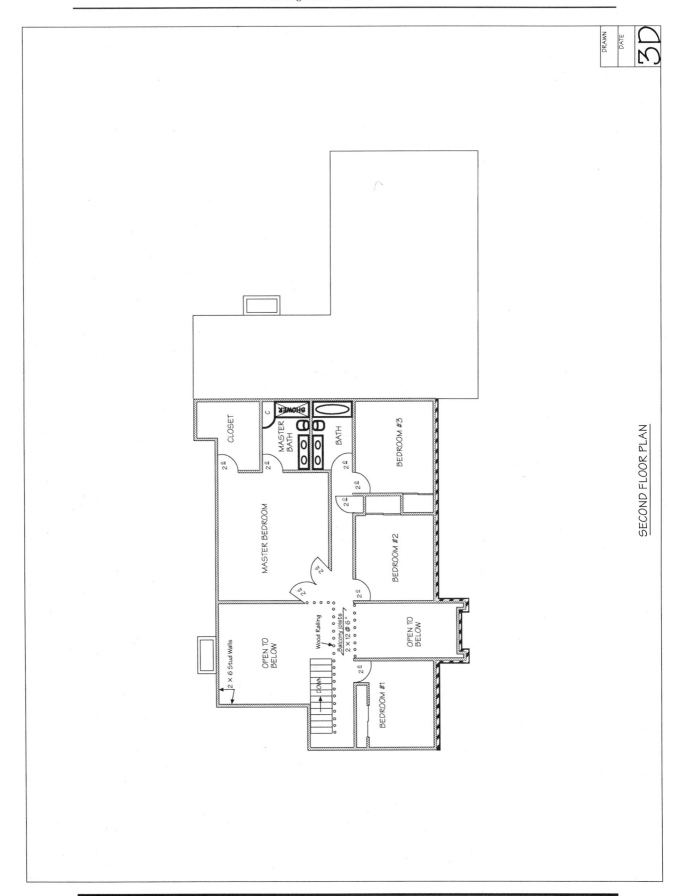

SECOND FLOOR PLAN

CLOSET

MASTER BATH

SHOWER

BATH

BEDROOM #3

MASTER BEDROOM

BEDROOM #2

OPEN TO BELOW

2 X 6 Stud Walls

Wood Railing

Balcony Joists
2 X 12 @ 6"

OPEN TO BELOW

DOWN

BEDROOM #1

DRAWN

DATE

3D

SECOND FLOOR PLAN

DRAWN

DATE

3F

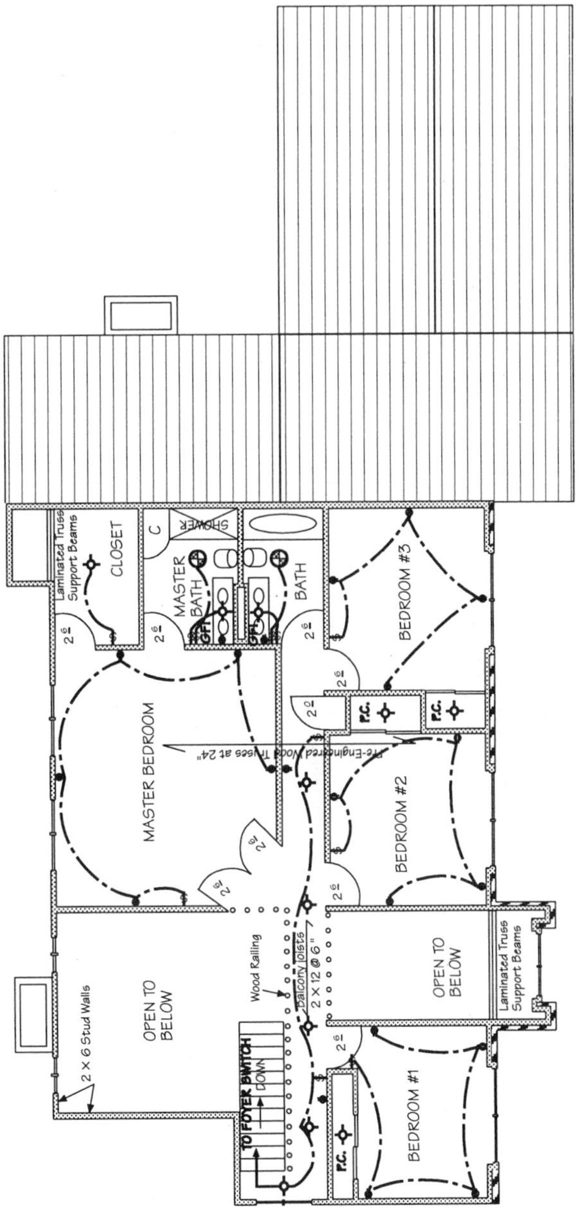

SECOND FLOOR PLAN

Laminated Truss
Support Beams

CLOSET

C

SHOWER

MASTER
BATH

BATH

BEDROOM #3

GFI

GFI

2⁰

2⁰

2⁰

2⁰

2⁰

P.C.

P.C.

MASTER BEDROOM

Pre-Engineered Wood Trusses at 24"

BEDROOM #2

2⁰

2⁰

2⁰

OPEN TO
BELOW

Laminated Truss
Support Beams

2 × 6 Stud Walls

OPEN TO
BELOW

Wood Railing

Balcony Joists

2 × 12 @ 6

TO FOYER SWITCH
DOWN

2⁰

BEDROOM #1

P.C.

131

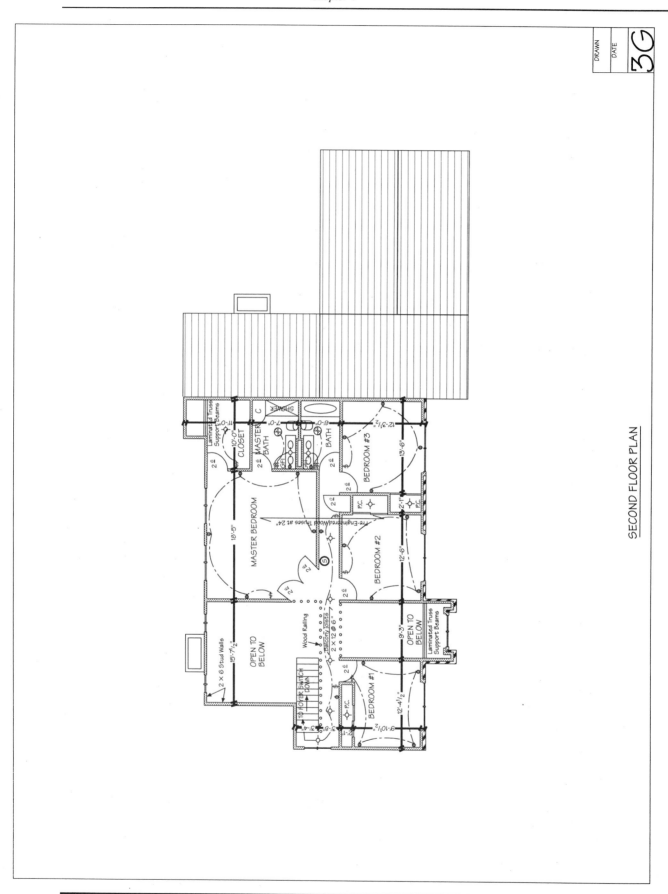

SECOND FLOOR PLAN

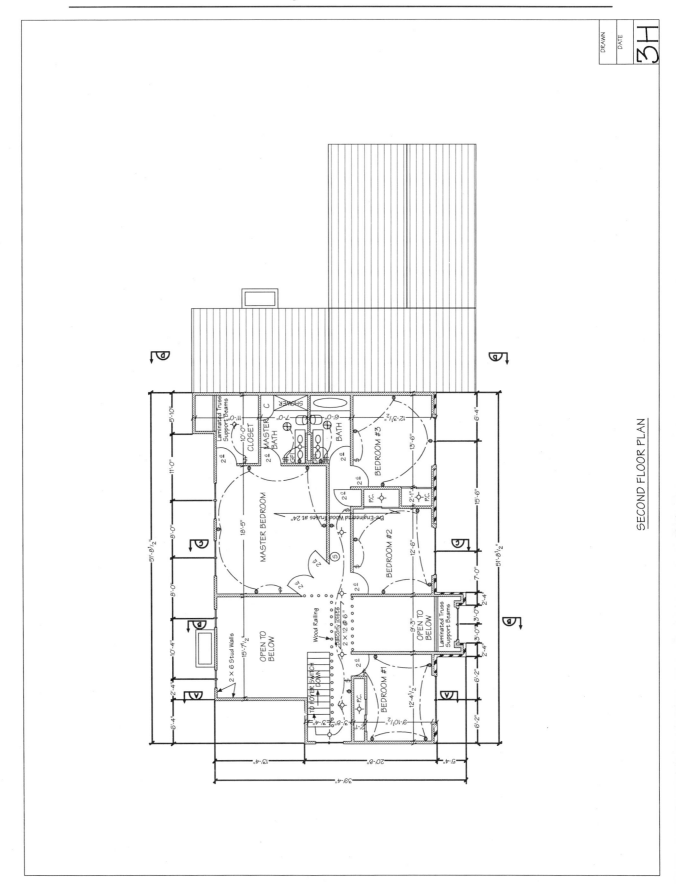

SECOND FLOOR PLAN

4—Sections

Drawing 4A—Foundation Walls

Drawing 4B—Slabs and Foundation Drain

Drawing 4C—First Story Floor

Drawing 4D—First Story Walls

Drawing 4E—Second Story Floors

Drawing 4F—Second Story Walls

Drawing 4G—Roof Trusses

Drawing 4H—Porch Roof and Stairway

Drawing 4I—Brick, Siding, and Roofing

Drawing 4J—Insulation

Drawing 4K—Drywall

Drawing 4L—Dimensions

Drawing 4A—Foundation

All the drawings in this section are taken from four crossections through the house. The crossections serve to clarify details that cannot be understood on the plan views (Drawings 1, 2, and 3).

Drawing 4A shows the foundation walls and footings. Note that the foundations go to different depths. The depth of a footing is controlled by many factors, including:

- Footing must be below frost level. When soil freezes, it tends to swell upwards. If the footing is above the freeze line, it will be pushed upward during the swelling causing damage to the house. If the footing is below the frost line, the swelling will not uplift it. The frost level varies from non-existent in the southern United States to several feet in the northern United States. A map shown in Figure 4-27 gives a very rough approximate depth of the frost levels.

- The footing must be supported on soil that can withstand the loads. Typically, the first few feet of soil are organic material or other unsuitable soils. The footing must extend below these levels to reach good soil.

- The foundation walls must extend low enough into the ground to provide the proper headroom in the basement.

Also shown in Drawing 4A is the gravel base for the basement and garage slabs.

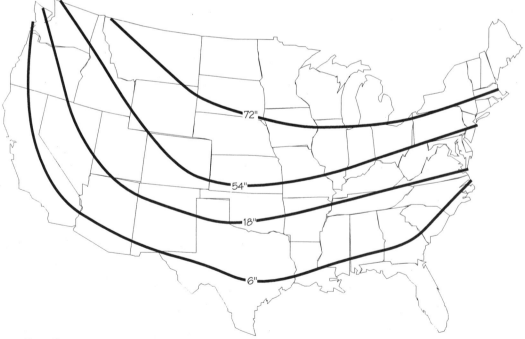

Note: This map is very general and shows an overview of frost depths. This drawing should not be used under any circumstances for determining the actual frost depth. It is only to give the reader a simplified view of frost depth.

Figure 4-27 Frost level

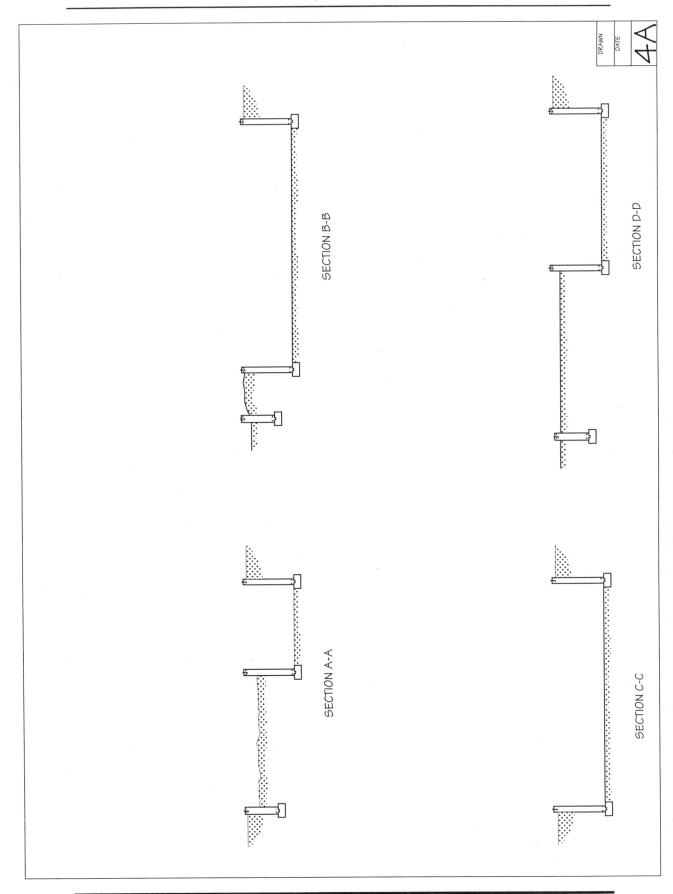

SECTION A-A

SECTION B-B

SECTION C-C

SECTION D-D

DRAWN

DATE

4A

Drawing 4B—Slabs and Foundation Drain

This drawing shows the concrete slabs in the garage and basement. These slabs are poured on a stone base. The ends of the slab are poured directly on the foundation wall footing as shown in the detail of Fig. 4-28.

As concrete dries and hardens, it shrinks in size. If the concrete can shrink freely, it does not experience any cracking. If the ends are restrained, the concrete slab cannot shrink freely and cracks develop in the slab. For this reason, a ⅜ inch thick piece of construction felt is placed between the side of the slab and the base of the wall. This material, called a bond breaker, prevents the two elements from bonding together and thereby prevents damaging restraint.

Piping, and the sump pits are embedded in the basement slab. The piping is used as a water drain should the basement ever flood from external sources or from overflow of the water heater. The piping must be installed prior to pouring the basement slab. A sump pit ring must also be embedded in the slab (Fig. 4-29).

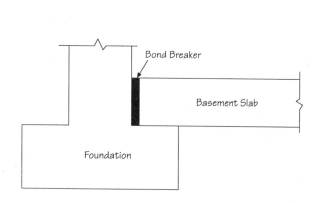

Figure 4-28 Wall-slab interface

Figure 4-29 Sump pit ring

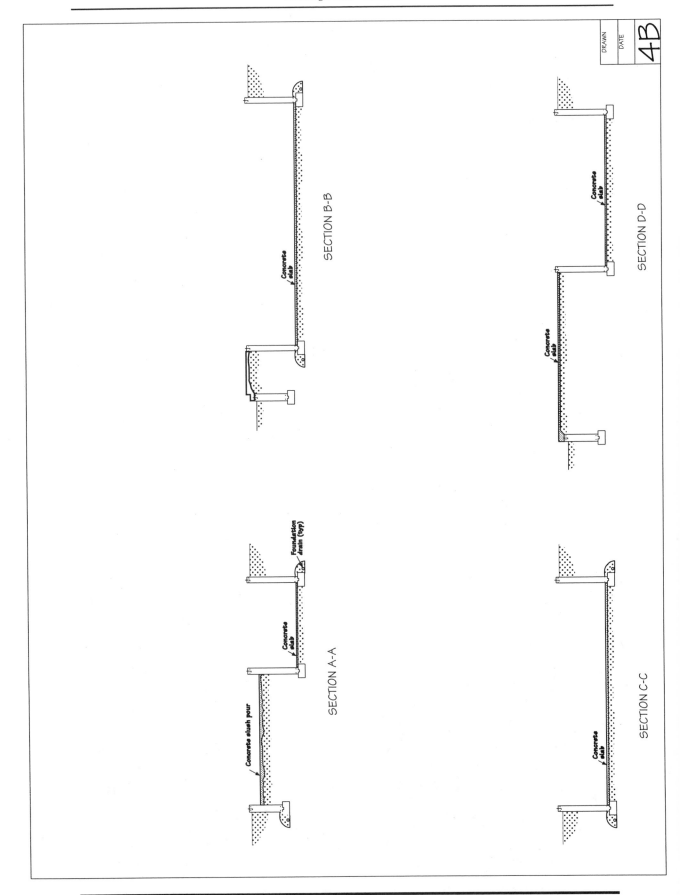

SECTION A-A

Concrete slush pour

Concrete slab

SECTION B-B

Concrete slab

Foundation drain (typ)

SECTION C-C

Concrete slab

SECTION D-D

Concrete slab

Concrete slab

DRAWN

DATE

4B

Drawing 4C—First Story Floor

The floor system consists of a wood joist and plywood deck as shown in Fig. 4-30. Bridging is required between the joists as shown in Fig. 4-31. Bridging is used to assist in distributing concentrated loads to adjacent joists so the floor acts as a unit rather than individual joists.

The joists, with the exception of the crawl space area, span from the foundation wall, across the steel beams to the opposite foundation wall. The joists are secured by nailing them to sill plates on the top of the wall and beam. The sill plate is a board laid on its side as shown in Figs. 4-32 and 4-33.

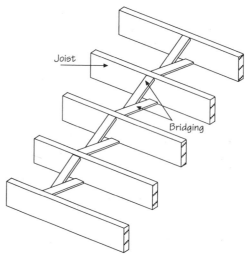

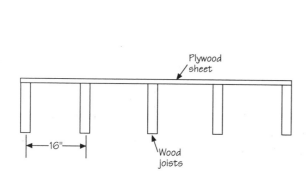

Figure 4-30 Floor system

Figure 4-31 Bridging

Figure 4-32 Sill plate on top of steel beam

Figure 4-33 Sill plate connection to foundation

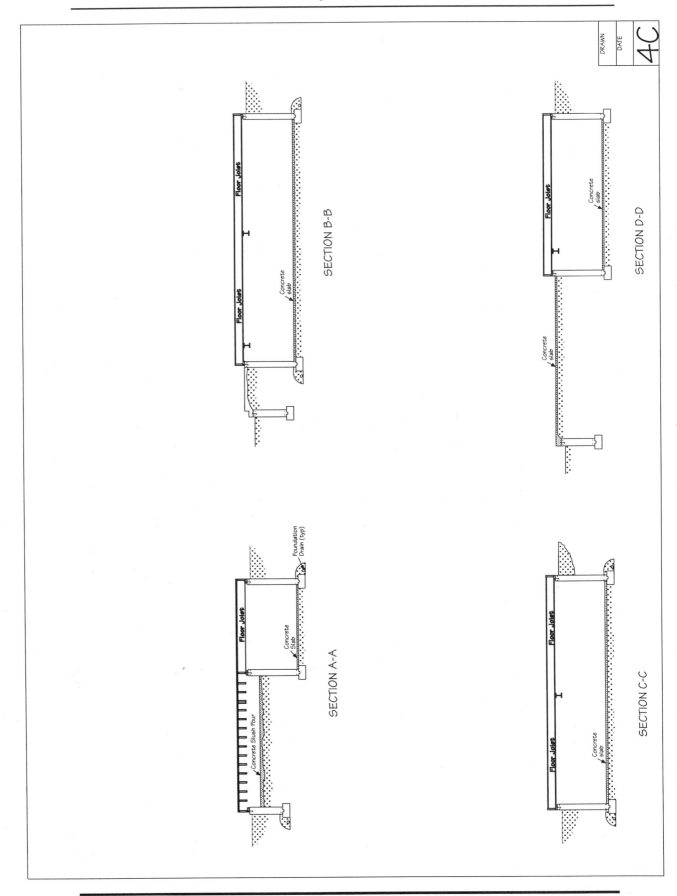

SECTION B-B

Floor Joist

Floor Joist

Concrete
Slab

SECTION A-A

Floor Joist

Concrete
Slab

Foundation
Drain (typ)

Concrete Slush Pour

SECTION D-D

Floor Joist

Concrete
slab

Concrete
slab

SECTION C-C

Floor Joist

Floor Joist

Concrete
slab

DRAWN

DATE

4C

Drawings 4D, 4E, and 4F—First Story Walls, Second Story Floors, and Second Story Walls

These topics were previously discussed. For details see:

Drawing 4D—See Drawing 2A and 2B

Drawing 4E—See Drawing 4C

Drawing 4F—See Drawing 2A and 2B

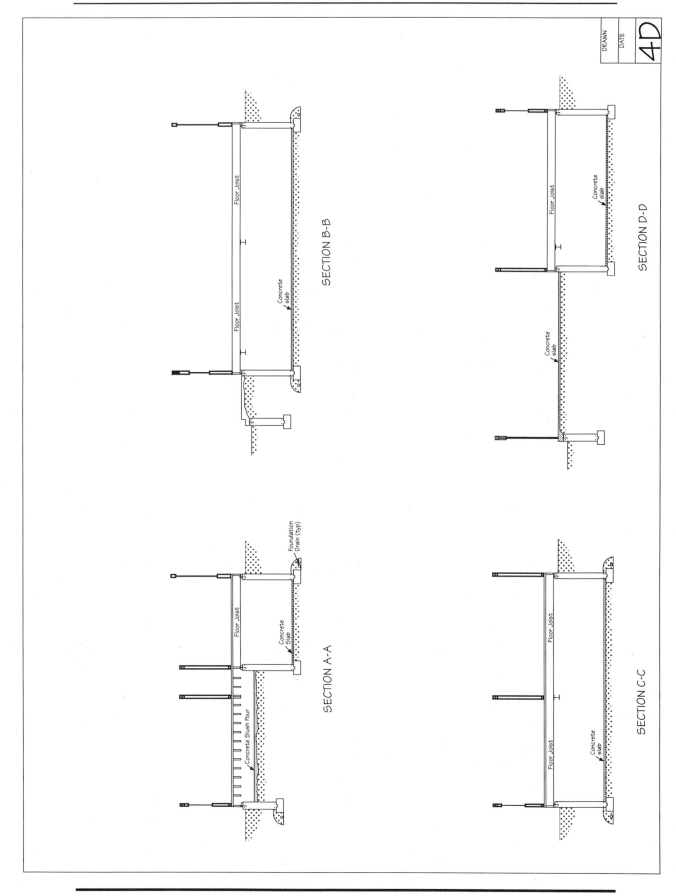

SECTION A-A

SECTION B-B

SECTION C-C

SECTION D-D

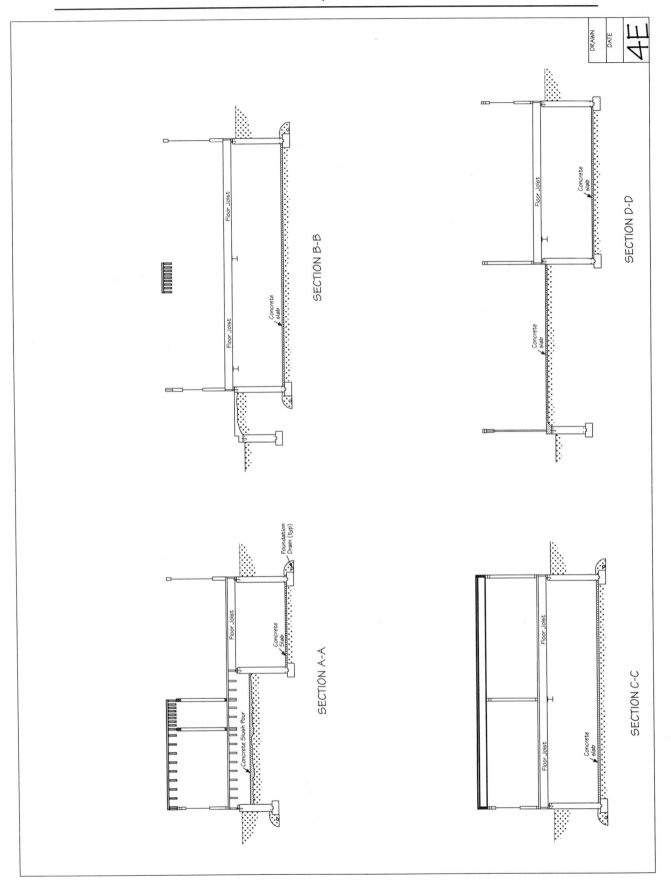

SECTION B-B

SECTION D-D

SECTION A-A

SECTION C-C

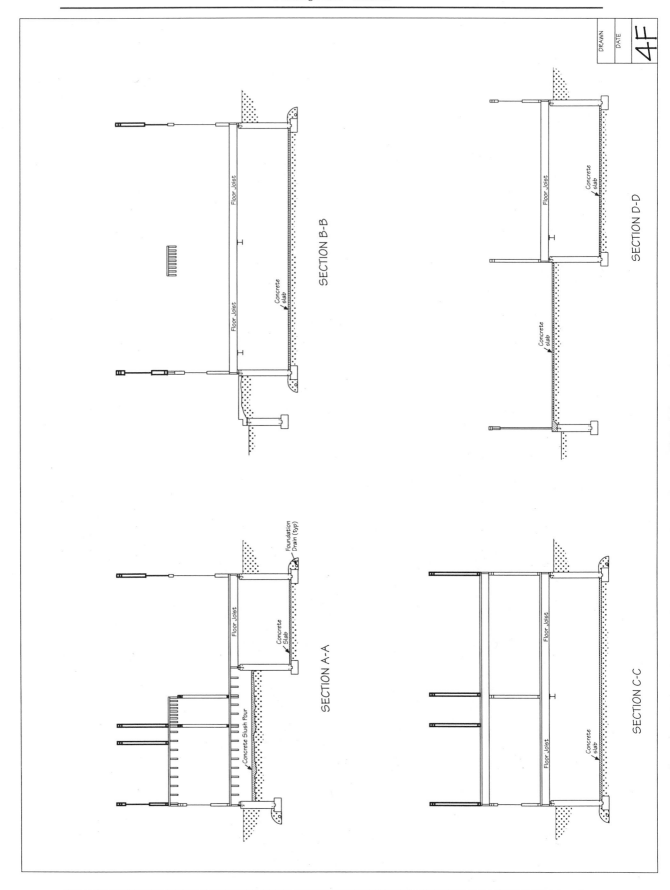

SECTION B-B

SECTION D-D

SECTION A-A

SECTION C-C

Foundation
Drain (typ)

Concrete
Slab

Concrete Slush Pour

Floor Joist

Concrete
slab

Floor Joist

Concrete
slab

Floor Joist

Floor Joist

Concrete
slab

DRAWN

DATE

4F

Drawing 4G and 4H—Roof Trusses, Porch Roof and Stairway

A truss is composed of wood members connected to other members to form a very efficient long span, load carrying system. Fig. 4-34 shows the joint of a truss. A joint is a location where individual members of the truss meet. The joint is held together by the connection plate shown in the figure.

The trusses that are shown in Drawing 4G are for purely schematic purposes. This style of truss are not the type that will be used, since that will be determined by the truss manufacturer. The terms used to describe the members of a truss are shown in Fig. 4-35. Truss members should not be altered at the construction site without the approval of the truss designer.

Trusses are designed to span from one exterior wall to the opposite exterior wall. The truss does not bear on interior walls. A gap should be provided between the interior walls and the bottom chord of the truss. If the truss is bearing on an interior wall and the truss is not designed for this condition, failure could occur as shown in Fig. 4-36.

The trusses overhang the exterior walls of the house. This overhang is called an eave (Fig. 4-37). The eave allows ventilation to enter the attic space. Improper ventilation can result in rotting of the truss members, mildew, and ice accumulation on the roof.

Figure 4-34 Joint of truss

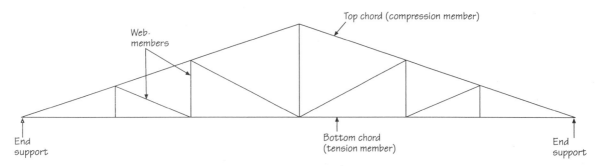

Figure 4-35 Truss members

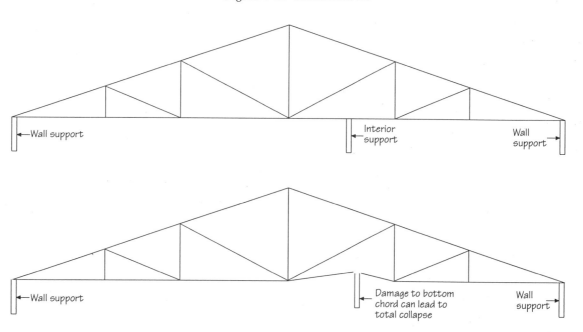

Figure 4-36 Distress in truss from bearing on interior walls

Figure 4-37 House eave

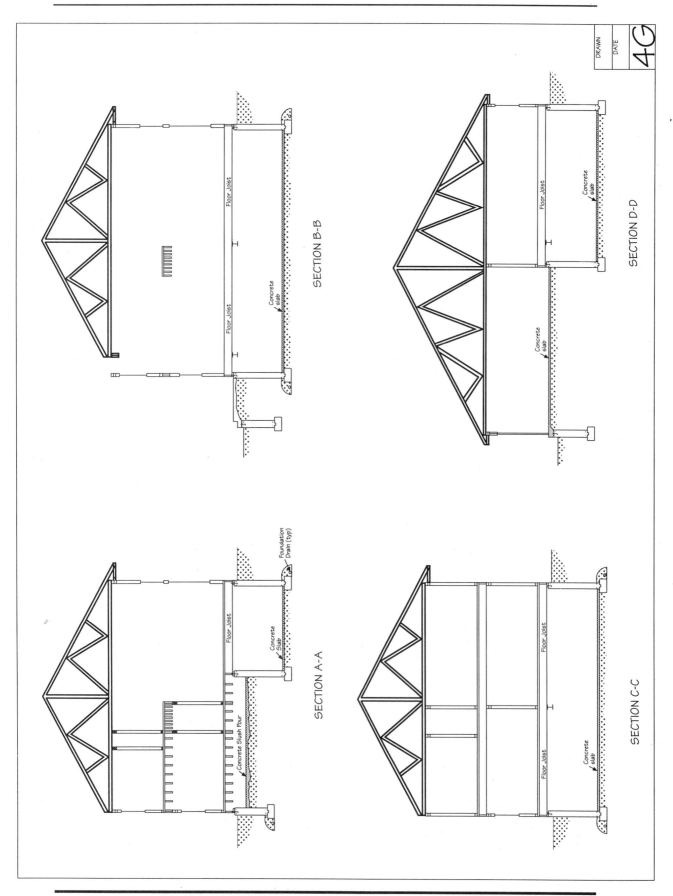

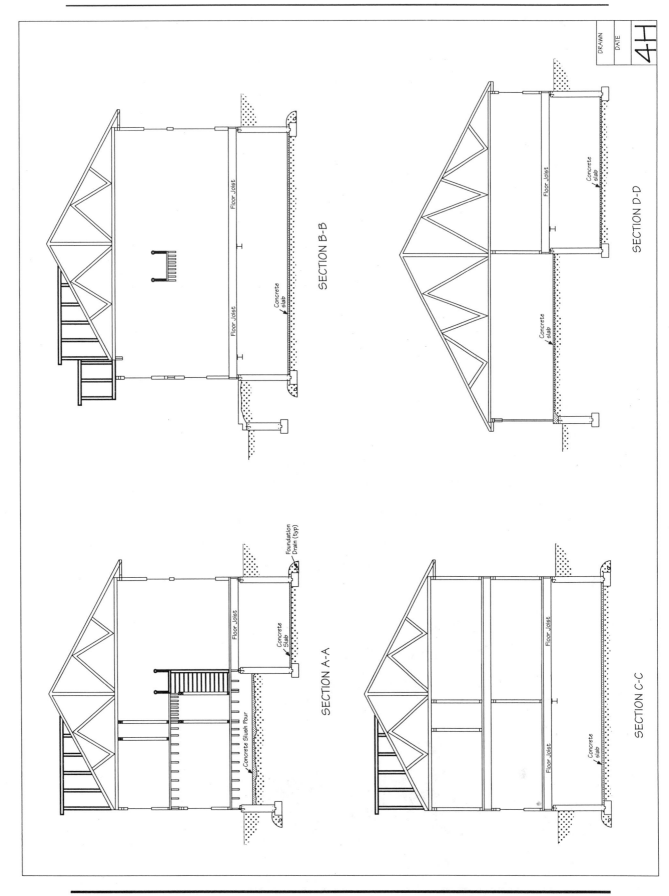

SECTION B-B

SECTION D-D

Floor Joist

Floor Joist

Concrete slab

Concrete slab

Floor Joist

SECTION A-A

SECTION C-C

Floor Joist

Concrete Slab

Foundation Drain (typ)

Concrete Slush Pour

Floor Joist

Floor Joist

Concrete slab

DRAWN

DATE

4H

Drawing 4I—Brick, Siding, and Roofing

Information on brick can be found by referring to Drawing 2F.

Wood siding is specified for the two sides and rear of the house. The four most popular sidings are:

Wood siding— usually cedar boards that can be placed either horizontal or vertical. Wood siding should have two coats of paint prior to installation and should be painted every five or so years.

Stucco— a concrete paste that is applied to vertical lath surfaces. Coloring can be added to the concrete mixture or it can be painted afterwards. This material can last many years without painting.

Aluminium— aluminum planks with a baked enamel paint finish. This material is maintenance free and need not be painted.

Vinyl— is very similar in looks and quality to aluminium siding.

Fig. 4-38 shows the installation of wood siding.

The roof must shed water without any of the water penetrating into the house. This is accomplished by having a pitched roof and a weather resistant covering. The weather resistant covering consists of felt paper with shingles on top of the roof plywood. The two most common types of shingles are:

Asphalt— usually a fiberglass and asphalt sheet with mineral grains embedded on the top surface.

Wood— these shingles are split from cedar boards and are called shakes.

Figure 4-38 Siding installation

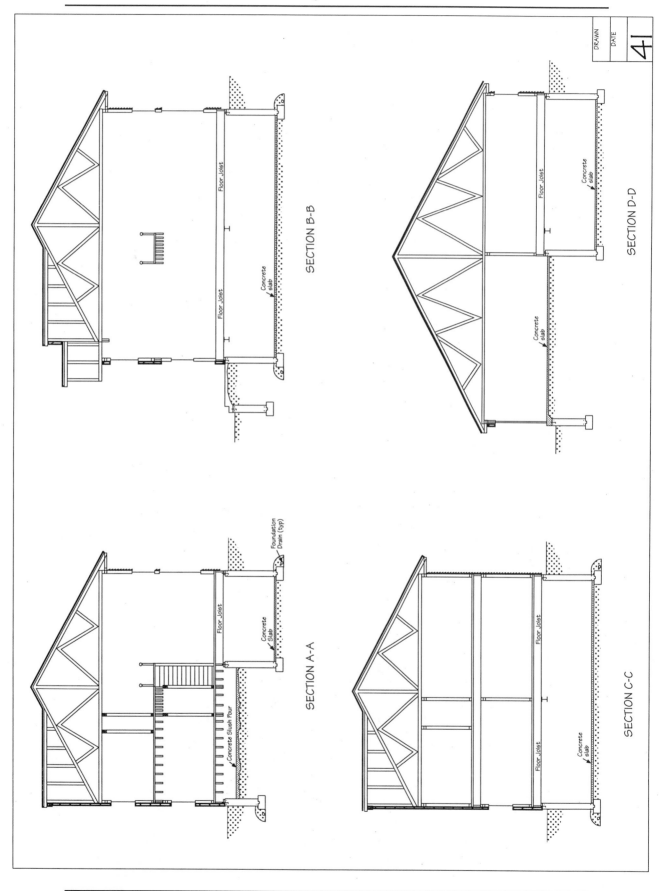

SECTION B-B

SECTION D-D

SECTION A-A

SECTION C-C

Floor Joist

Concrete slab

Concrete Slush Pour

Foundation Drain (typ)

Floor Joist

Concrete Slab

Floor Joist

Concrete slab

Concrete slab

Floor Joist

Drawing 4J and 4K—Insulation and Drywall

Insulation is needed to keep in the heat in the winter and the cool air in the summer. The insulation can be from rolled sheets (Fig. 4-39) or can be loose material (Fig. 4-40).

The effectiveness of insulation is measured by the "R-number." The greater the value of R, the more efficient the insulation. The required value of R is a function of where the house is located. The typical values of R are given below.

Ceilings	Far Northern U.S. and mountain areas	38
	Middle U.S.	30 to 33
	Lower U.S.	26
Walls	Far Southern U.S.	13
	All others	19
Floor	Upper half of the U.S.	19 to 22
	Lower half of the U.S.	11 to 13

Note that these figures are not applicable to the extreme West Coast.

The R value is related to the type of insulation as well as its thickness. For fiberglass rolled sheets, the required thickness is approximately equal to the R value divided by 3. For loose material fiberglass insulation, the R value divided by 2 gives an estimate of the required thickness.

The cold and warm air interface of the house produces condensation. To prevent the moisture problem, a vapor barrier is attached to the inside face of the insulation. A vapor barrier is foil or paper material that resists the infiltration of moisture.

Drywall is the interior finish, made of preformed gypsum boards, placed over the stud walls. Drywall is typically ½ inch, but may need to be thicker if used in fire resistant walls.

Figure 4-39 Rolled sheet insulation

Figure 4-40 Loose insulation

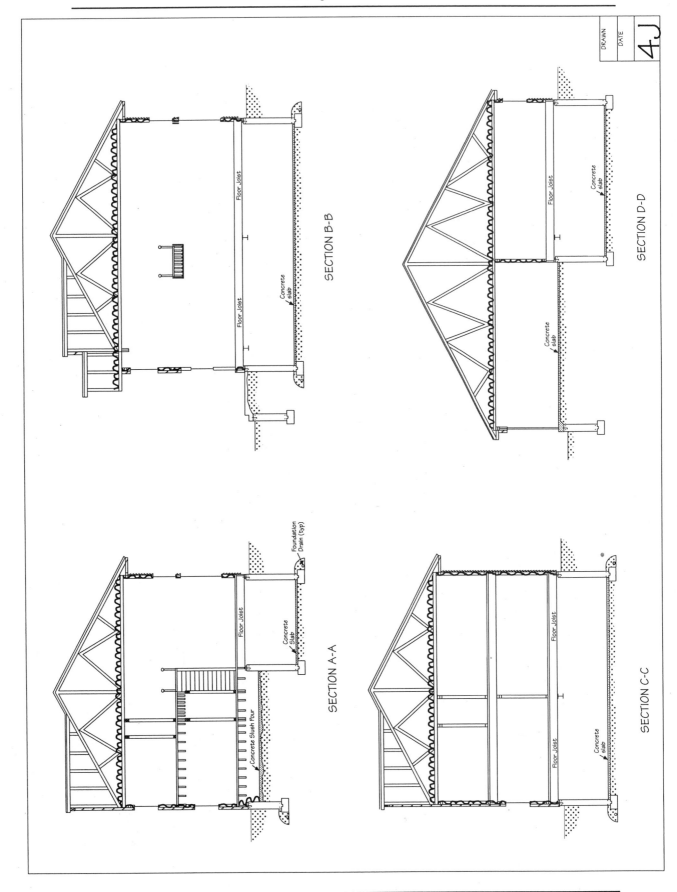

SECTION B-B

Floor Joist

Floor Joist

Concrete
slab

SECTION D-D

Floor Joist

Concrete
slab

Concrete
slab

SECTION A-A

Floor Joist

Concrete
Slab

Concrete Slush Pour

Foundation
Drain (typ)

SECTION C-C

Floor Joist

Floor Joist

Concrete
slab

DRAWN

DATE

4J

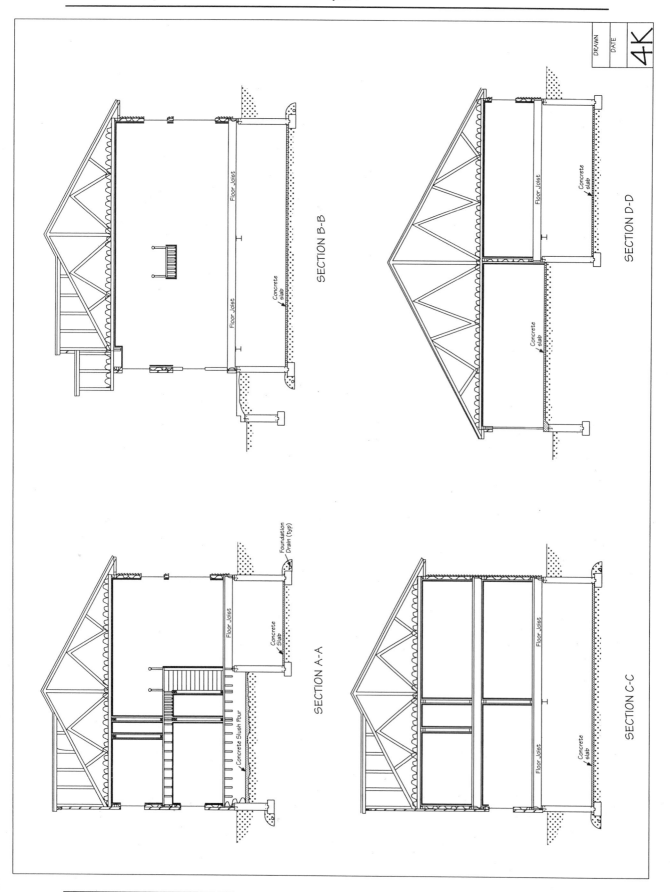

SECTION B-B

SECTION D-D

SECTION A-A

SECTION C-C

Drawing 4I—Dimensions

Dimensioning is self-explanatory.

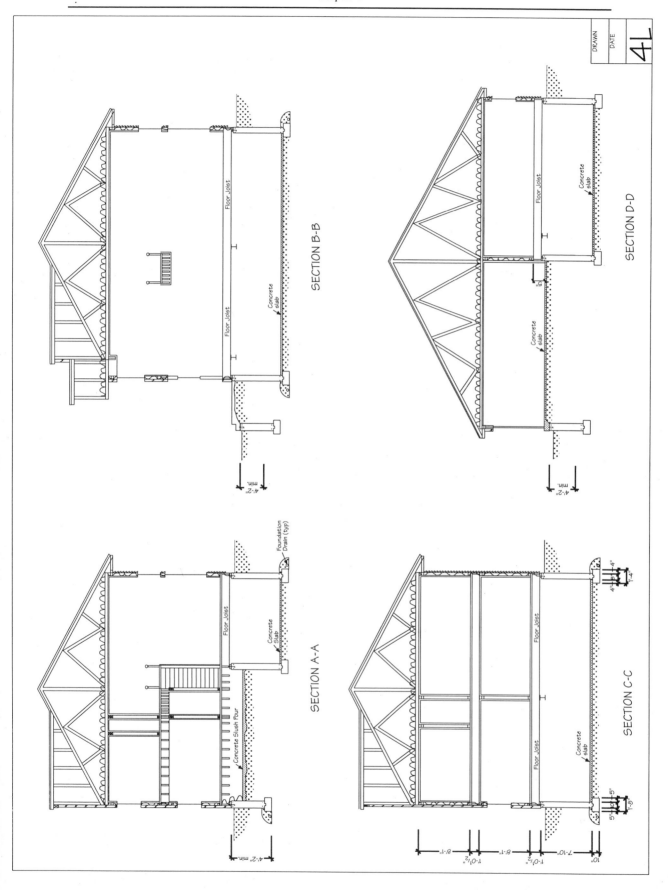

SECTION B-B

SECTION A-A

SECTION D-D

SECTION C-C

Floor Joist

Concrete slab

Concrete Slab

Foundation Drain (typ)

Concrete Slush Pour

Floor Joist

Floor Joist

Floor Joist

Floor Joist

Concrete slab

Concrete slab

4'-2" min.

4'-2" min.

4'-2" min.

DRAWN

DATE

4L

5, 6—Front, Rear, and Side Elevations

These drawings are self-explanatory.

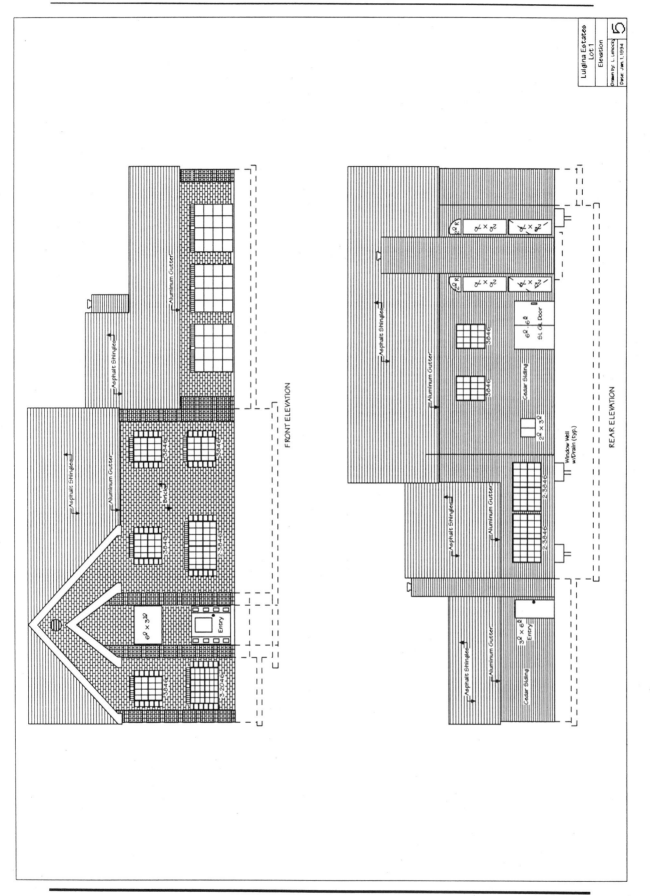

FRONT ELEVATION

REAR ELEVATION

Luigina Estates
Lot 1
Elevation
Drawn by: L. Lenocky
Date: Jan. 1, 1994

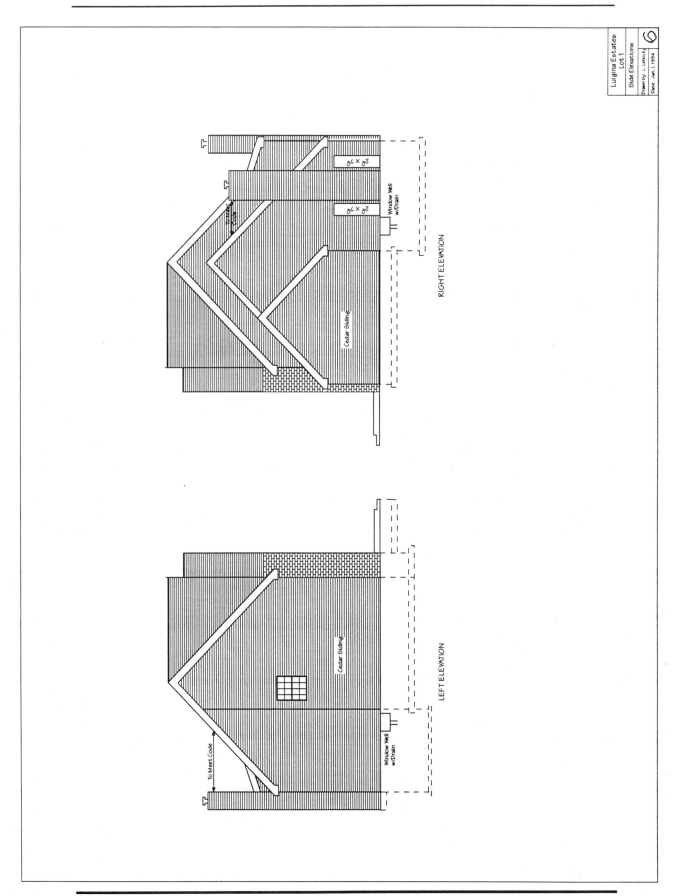

7—Details

Drawing 7A and 7B—Roof Slopes, Lighting and Ventilation

The slope of the roof and the direction of the slope is shown in the plan view of Drawing 7A. This roof slope information is used for:

- Determination of the quantity of roofing material required.

- Providing necessary information to the fabricator of the trusses.

- Helping to clarify other drawings.

Figures 4-41 through 4-43 show the shapes of various types of roofs.

Drawing 7B provides a table giving the required and provided ventilation and lighting.

The amount of ventilation and lighting required is mandated by the applicable building code. Refer to Drawing 2G.

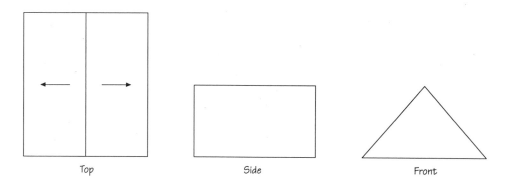

Figure 4-41 Gamble roof

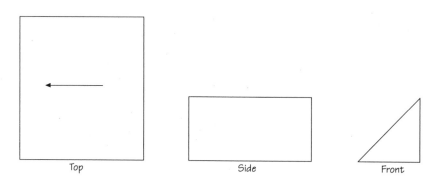

Figure 4-42 Shed roof

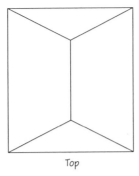
Top

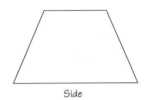
Side

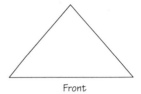
Front

Figure 4-43 Hip roof

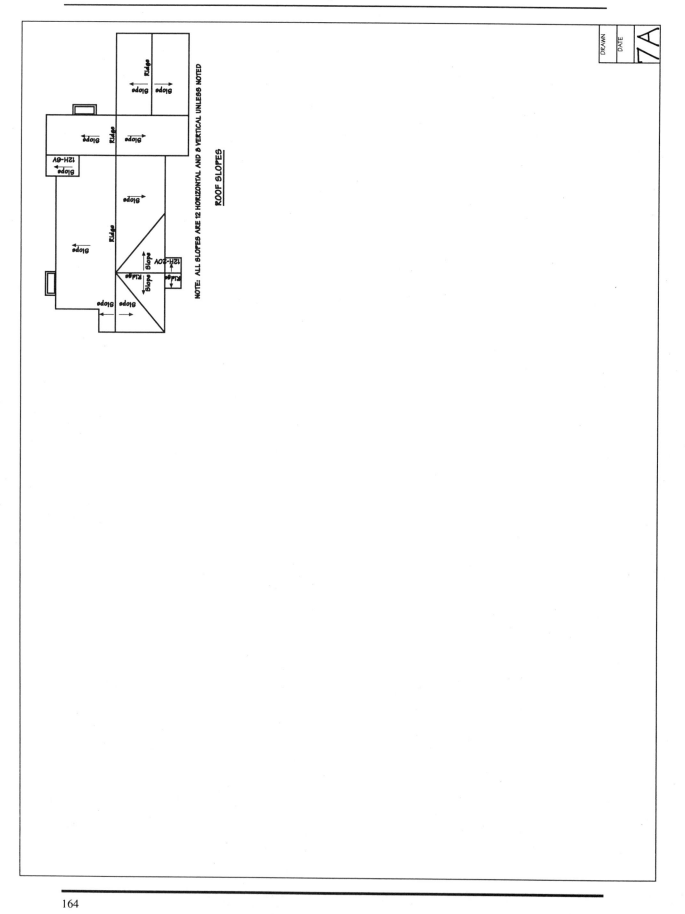

ROOF SLOPES

NOTE: ALL SLOPES ARE 12 HORIZONTAL AND 6 VERTICAL UNLESS NOTED

7A

DRAWN

DATE

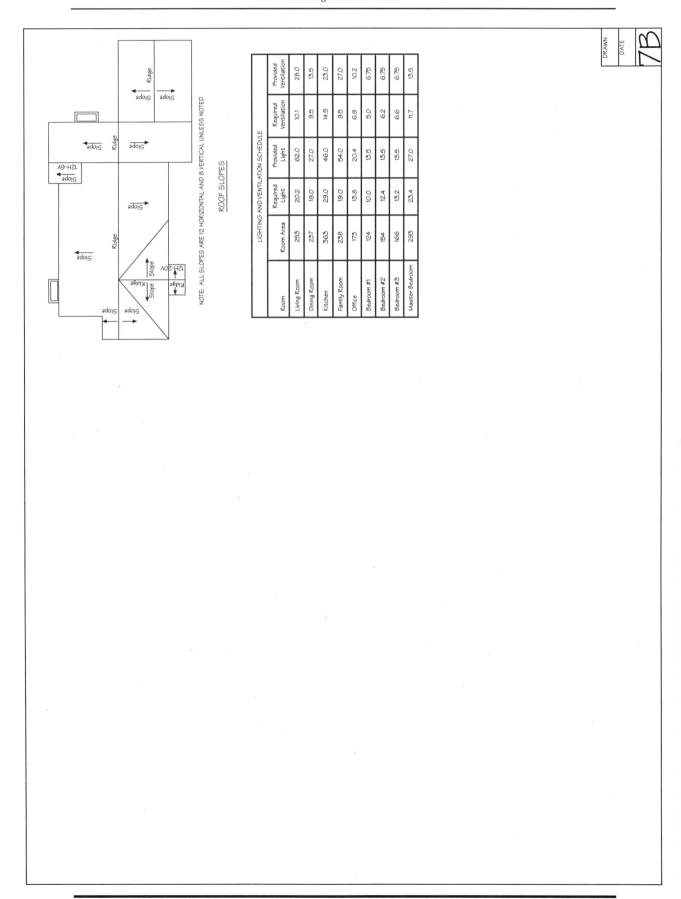

ROOF SLOPES

NOTE: ALL SLOPES ARE 12 HORIZONTAL AND 8 VERTICAL UNLESS NOTED

LIGHTING AND VENTILATION SCHEDULE

Room	Room Area	Required Light	Provided Light	Required Ventilation	Provided Ventilation
Living Room	253	20.2	62.0	10.1	28.0
Dining Room	237	19.0	27.0	9.5	13.5
Kitchen	363	29.0	46.0	14.5	23.0
Family Room	238	19.0	54.0	9.5	27.0
Office	173	13.8	20.4	6.9	10.2
Bedroom #1	124	10.0	13.5	5.0	6.75
Bedroom #2	154	12.4	13.5	6.2	6.75
Bedroom #3	166	13.2	13.5	6.6	6.75
Master Bedroom	293	23.4	27.0	11.7	13.5

DRAWN

DATE

7B

Drawing 7C, 7D, and 7E—Foundation, 1st and 2nd Floors, and Roof

These drawings provide additional details for wall and foundation construction details. These drawings are a supplement to the details in Drawing 4. The additional details provided in these drawings are:

- Information on what is located under the siding and brick.

- Details on the floor system.

- Details for basement slab (thickness, reinforcing, and base material).

- Keyway size.

- Specification for sill plate material and attachment.

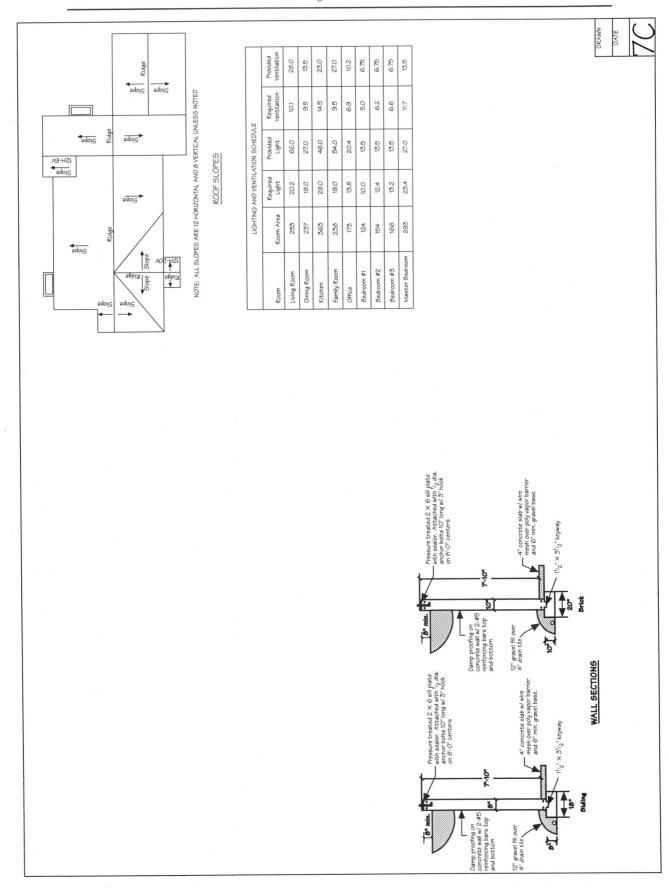

ROOF SLOPES

NOTE: ALL SLOPES ARE 12 HORIZONTAL AND 8 VERTICAL UNLESS NOTED

LIGHTING AND VENTILATION SCHEDULE

Room	Room Area	Required Light	Provided Light	Required Ventilation	Provided Ventilation
Living Room	253	20.2	62.0	10.1	28.0
Dining Room	237	19.0	27.0	9.5	13.5
Kitchen	363	29.0	46.0	14.5	23.0
Family Room	238	19.0	54.0	9.5	27.0
Office	173	13.8	20.4	6.9	10.2
Bedroom #1	124	10.0	13.5	5.0	6.75
Bedroom #2	154	12.4	13.5	6.2	6.75
Bedroom #3	166	13.2	13.5	6.6	6.75
Master Bedroom	293	23.4	27.0	11.7	13.5

WALL SECTIONS

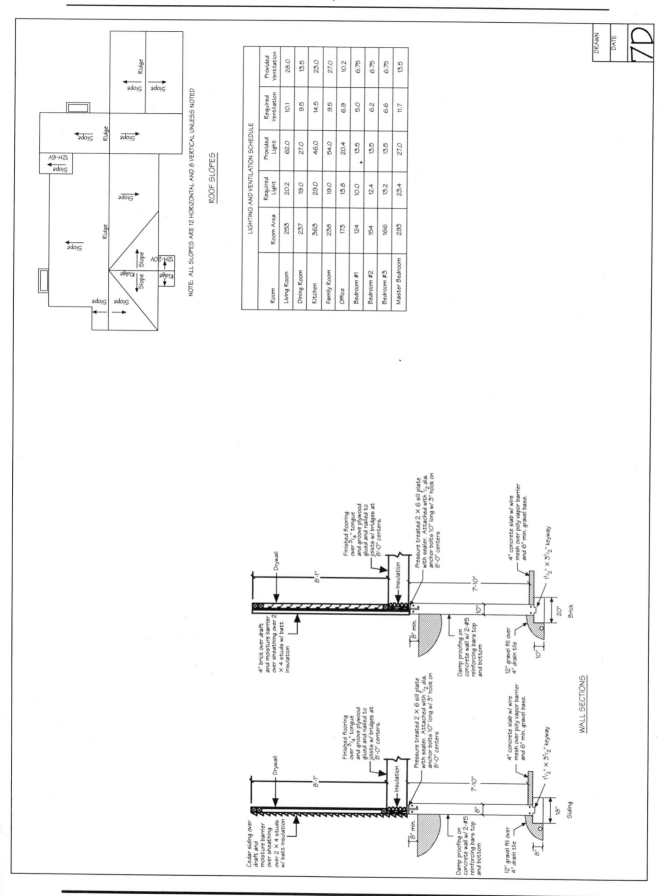

NOTE: ALL SLOPES ARE 12 HORIZONTAL AND 8 VERTICAL UNLESS NOTED

ROOF SLOPES

LIGHTING AND VENTILATION SCHEDULE

Room	Room Area	Required Light	Provided Light	Required Ventilation	Provided Ventilation
Living Room	253	20.2	62.0	10.1	28.0
Dining Room	237	19.0	27.0	9.5	13.5
Kitchen	363	29.0	46.0	14.5	23.0
Family Room	238	19.0	54.0	9.5	27.0
Office	173	13.8	20.4	6.9	10.2
Bedroom #1	124	10.0	13.5	5.0	6.75
Bedroom #2	154	12.4	13.5	6.2	6.75
Bedroom #3	166	13.2	13.5	6.6	6.75
Master Bedroom	293	23.4	27.0	11.7	13.5

WALL SECTIONS

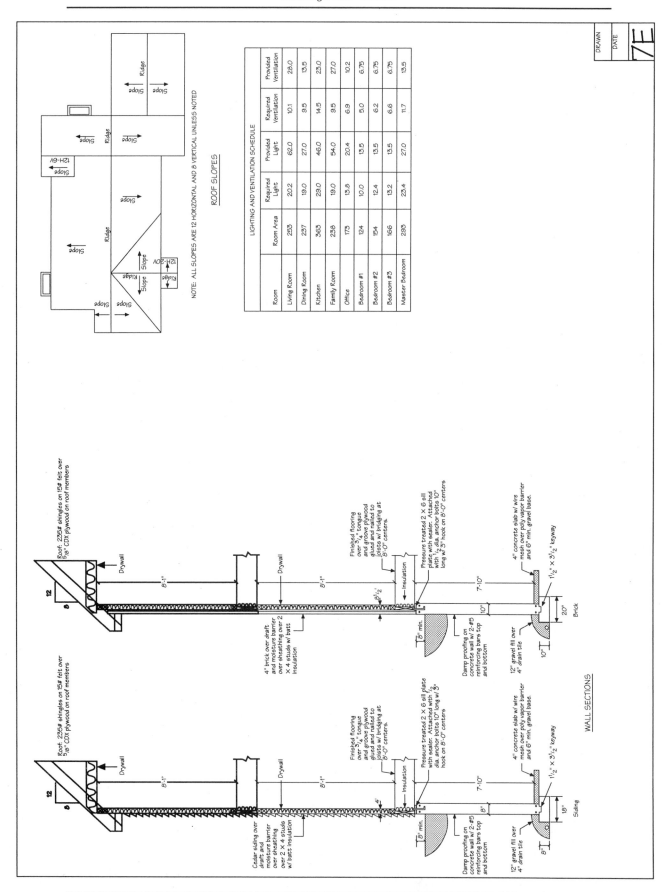

ROOF SLOPES

NOTE: ALL SLOPES ARE 12 HORIZONTAL AND 8 VERTICAL UNLESS NOTED

LIGHTING AND VENTILATION SCHEDULE

Room	Room Area	Required Light	Provided Light	Required Ventilation	Provided Ventilation
Living Room	253	20.2	62.0	10.1	28.0
Dining Room	237	19.0	27.0	9.5	13.5
Kitchen	363	29.0	46.0	14.5	23.0
Family Room	238	19.0	54.0	9.5	27.0
Office	173	13.8	20.4	6.9	10.2
Bedroom #1	124	10.0	13.5	5.0	6.75
Bedroom #2	154	12.4	13.5	6.2	6.75
Bedroom #3	166	13.2	13.5	6.6	6.75
Master Bedroom	293	23.4	27.0	11.7	13.5

WALL SECTIONS

169

Drawing 7F—Overhang Ventilation Detail

The roof space must be properly vented to prevent a variety of roof problems, including rotting members, ice build up, and mildew. Ventilation is achieved by air flowing in through the overhang detail and flowing out through root vents as shown in Fig. 4-44.

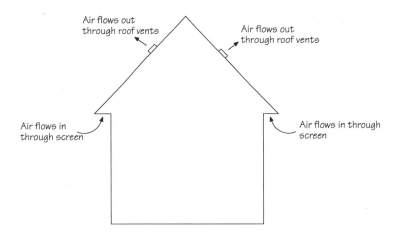

Figure 4-44 Ventilation of attic space

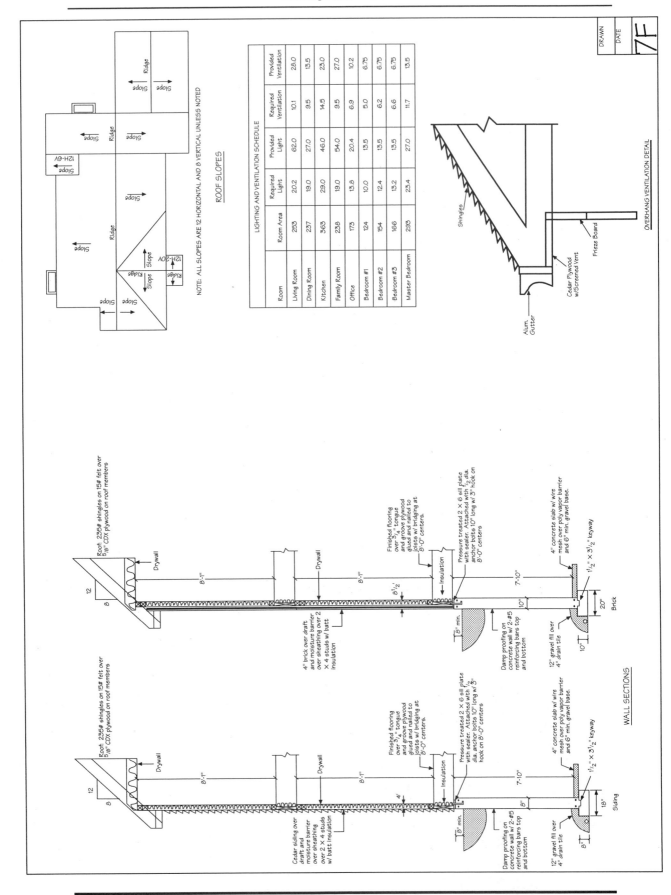

ROOF SLOPES

NOTE: ALL SLOPES ARE 12 HORIZONTAL AND 8 VERTICAL UNLESS NOTED

LIGHTING AND VENTILATION SCHEDULE

Room	Room Area	Required Light	Provided Light	Required Ventilation	Provided Ventilation
Living Room	253	20.2	62.0	10.1	28.0
Dining Room	237	19.0	27.0	9.5	13.5
Kitchen	363	29.0	46.0	14.5	23.0
Family Room	238	19.0	54.0	9.5	27.0
Office	175	13.8	20.4	6.9	10.2
Bedroom #1	124	10.0	13.5	5.0	6.75
Bedroom #2	154	12.4	13.5	6.2	6.75
Bedroom #3	166	13.2	13.5	6.6	6.75
Master Bedroom	293	23.4	27.0	11.7	13.5

OVERHANG VENTILATION DETAIL

WALL SECTIONS

Chapter 5

Specifications

Design drawings are the most important part of construction documents. As seen in the previous chapter, the design drawings provide the necessary dimensions, member sizes, as well as other essential information. But even the most detailed drawings cannot list all the information needed to construct a house. This is the reason specifications are needed to supplement design drawings. Specifications are written instructions that provide information on details not explained on the drawings.

Specifications for construction projects range from tens of thousands of pages for high-rise buildings to five to ten pages for residential construction. This chapter presents items (located within the rectangles) typically found in residential construction specifications, as well as explanations of their significance. Forms to aid in organization of material selection are provided at the end of this chapter. The first form is organized by location; the second is organized by material.

---

(1) Excavation—The contractor shall perform the following excavation at the contractors cost:

- Removal of tree or bush growth in area of excavation.
- Strip topsoil and stockpile.
- All necessary foundation excavating.
- All necessary foundation backfilling.
- Spreading of stockpiled soil upon completion of the structure.

The purchaser shall pay the expense for any labor or material for the following items:

- Removal of trees or bush growth outside excavation area
- Additional grading beyond spreading of topsoil.
- Removal and disposal of any buried items.

A significant amount of litigation occurs in all types of building and bridge construction as a result of problems related to excavation. Digging can unearth a multitude of problems, including:

- Improperly located utility lines.
- Boulders.

- Extremely saturated soils.

- Buried hazardous materials.

These are just a sample of potential problems. The contract must list work to be performed by the excavator, as well as placing the burden on removal of buried items on the purchaser.

> (2) Concrete Foundations—All footings and foundations shall reach at least 3000 psi in 28 days. Steel reinforcing bars shall be ⅜ inch diameter and installed at locations shown on design drawings.

The strength of concrete at 28 days is the industry standard for comparison. A compressive strength of 3000 psi is an easily obtained as well as a relatively low strength. The strength of the hardened concrete is a function of the relative amounts of water and cement in the mixture and is determined by the ready-mix concrete supplier. Steel reinforcing bars are needed for crack control if shown on the design drawings. Reinforcing bars are typically grade 60 and are designated by a number which is related to their diameter (i.e., ⅜ is a #3 bar, ⅝ is a #5 bar).

> (3) Steel Members—Steel pipe columns shall be 3½ inches outside diameter and filled with concrete. The steel pipe columns shall conform to ASTM standards.
>
> Steel beams shall be W8 × 21, or as specified on the design drawings. Steel beams shall conform to ASTM standards.

Steel members are the sizes listed above and meet the American Society of Testing Material (ASTM) standards.

> (4) Foundation Wall Dampproofing—All portions of the foundation wall that will be buried below grade shall have a coating of an asphalt based foundation dampproofing on the exterior face. Dampproofing shall be applied when the temperature is above 50°F and no rain or snow will occur within 24 hours. Dampproofing shall consist of two layers applied in opposite directions.

Water penetration through the basement walls is inhibited by the application of an asphalt based coating. The coating may be brushed or sprayed on. Note that there is a difference between "dampproofing" and "waterproofing". Waterproofing prohibits rather than inhibits the penetration of water and is much costlier to provide.

(5) Footing Drains—Footing drains shall be a minimum of 4 inch diameter plastic perforated pipe. The pipe shall meet ASTM standards. The drain shall be placed in 12 inches of stone. The footing drain shall empty by gravity away from the structure or into a sump pit.

The footing drain pipe is buried in stone to provide good drainage at that location.

(6) Fireplace—Fireplaces shall be prefabricated with the size shown on the design drawings. Fireplace design shall be certified by the American Gas Association and have appropriate venting.

If the fireplace was not prefabricated, additional information such as fireplace lining and type of construction would be included in this part of the specifications.

(7) Exterior Walls—Exterior walls shall be constructed of hem fir 2×12 studs unless noted differently on the drawing. Exterior cladding shall be cedar siding or brick over wood sheathing as shown on the design drawings. A building wrap shall be used between the cladding and the sheathing. Cedar siding shall be stained with two coats.

(8) Interior Walls—Interior walls shall be constructed of hem fir 2×4 studs that are 16 inches on center.

(9) Floor and Ceiling Framing—All wood framing shall be 2×8, 2×10, 2×12 and spaced as noted on the drawings. Wood shall be hem fir. Joists shall have bracing located between adjacent joists.

(10) Prefabricated Wood Trusses—Prefabricated trusses shall be designed to carry the appropriate loads. A stamp by a licensed structural engineer shall accompany each truss.

Items 7 through 10 cover the structural framing for the house. It is important to state in the specifications the species of wood to be used.

(11) Roofing—Shingles shall be 280 lbs. fiberglass with a one year warranty on labor and a minimum 25 year warranty on the material. Felt under shingles shall be a minimum of 15 pounds. The roofing material shall be secured to ¾ inch thick exterior grade plywood.

Roofing shingles are classified by weight. The weight specified is the weight of the shingles to cover 1000 square feet of roof. The greater the weight, the longer the life expectancy of the shingles. The plywood on top of the trusses must be sufficient to withstand exterior exposure.

> (12) Gutters and Downspouts—All gutters and downspouts shall be 4 inch aluminum box type construction.

Gutters and downspouts should be sized to adequately carry water runoff. The size is typically picked by experience based on other successful installations.

> (13) Windows—All windows shall be casement with aluminum-clad exterior and shall be manufactured by ABC windows or an approved equivalent. Window shall include screen and storm windows.

Windows can have a wide variation in quality. Problems can be avoided by listing at least the manufacturer and possibly even the model numbers.

> (14) Garage Doors—All garage doors shall be plain raised redwood panel doors. Redwood doors shall be weather treated.

The material of the garage doors should be specified. If the material is steel, then an insulated door should be specified in colder climates.

> (15) Septic System—Complete septic shall be installed as per plans, including stone, filter paper, pipe lines, septic tanks, and backfilling. All work shall meet county approval.

This item is only needed if an individual sewage disposal system is required. Connection to a public system requires excavation, furnishing, installing, and connecting a sewage pipe from the house to the trunk line of the public sewer system.

The septic system must be designed by a competent designer. The designer will size the tanks and pipelines to provide a system able to percolate the wastewater into the soil.

> (16) Driveway—The driveway shall consist of a 10 inch stone base below 4 inch thick asphalt. The stone shall be compacted prior to placing the asphalt.

The driveway can be either asphalt or concrete. A concrete driveway has a specification similar to that for the concrete slabs presented in the next item.

> (17) Concrete Slabs—All concrete slabs shall be poured on a minimum of 4 inches of gravel and visqueen. The concrete slabs shall be reinforced with welded wire fabric. The concrete mix shall be such that the strength of the concrete is at least 3500 psi within 28 days.

The slabs are poured on stone to provide good drainage and prevent settlement. The visqueen, which is a sheet of plastic, is placed in the stone layer to prevent the migration of moisture from the soil up through the slab. Welded wire fabric (WWF) is placed in the upper third of the slab to restrain surface cracks in the slab.

> (18) Plumbing—The general contractor shall provide the following:
>
> Water:
> - Piping for all sinks, bathrooms, dishwasher, water heaters, and all other items that require water.
> - 50 gallon hot water heater
> - Well
> - Kitchen sink
> - Utility sink
> - Shower bases
> - Faucets
> - Exterior faucet
> - Sump pumps
> - Water softener
>
> Wastewater:
> - Piping from all sinks, bathrooms, etc.
> - Septic system
>
> Gas:
> - Gas piping from exterior source to furnaces, stoves, fireplace log lighter, dryer, and water heater.
>
> Allowances:
> | Septic | $X,XXX |
> |---|---|
> | Water softener | $X,XXX |
> | Well | $X,XXX |
> | Plumbing fixtures | $X,XXX |

Most people associate plumbing with water supply and wastewater removal. The plumbing work also includes installing the sinks, hot water heater and sump pumps.

Certain items cannot be adequately estimated prior to construction. It is hard to estimate how deep a well will be until the drilling has started. For this type of item, the general contractor gives the purchaser an allowance. The allowance is the average price for the work and is just a gauge for costs. After the work item is completed and the final cost determined, the price is adjusted accordingly.

(19) Heating—Two high-efficiency gas burning furnaces with galvanized metal supply ducts shall be sized and installed. The furnace shall be ABC brand or of at least equal quality. Furnace shall be operable by digital thermometer with programming capability.

It is more efficient in some cases to have two separate furnaces for the first and second floor. This raises the initial cost of the house, but these costs will be recovered from the efficiency of the two furnace system. The specifications should clarify whether two furnaces will be provided.

The gas company provides service to the house at no cost to a location less than a specified length (i.e. 50 feet from roadway). If the purchaser would like to move the gas meter to a less conspicuous place or if the house is set back far from the front of the lot, the purchaser must bear the extra cost to extend the gas lines.

(20) Insulation—Insulation with the following R number shall be provided.

Ceiling   R = 33
Walls   R = 19
Floor   R = 22

Insulation shall only be provided in garage walls common to living areas. The exterior of the house shall be wrapped with draft reducing plastic.

The amount and efficiency of the insulation system effects not only the heating and cooling costs, but also the comfort level to the inhabitants. The "R" number, the measure of the efficiency of the insulation, should be clearly stated in the specifications to avoid substandard insulation.

(21) Paint—Two coats of flat latex paint shall be applied to all walls and ceiling. Colors other than white shall result in an extra charge.

An extra cost is included if a color other than white is used because it requires extra cleaning of equipment for the painting subcontractor.

(22) Trim—All interior doors and trim shall be pine. Doors shall be stained and sealed and of the six-panel style. Base trim and door casings as well as any crown moldings specified on drawings shall be provided by the contractor and shall be stained and sealed.

Trim costs can vary widely depending on the complexity of the installation and the type of materials. If the trim is nonstandard, it should be explained in detail in the specifications or drawings. It should be noted whether the trim will be pine or oak.

---

(23) Cabinets—Kitchen cabinets shall be constructed of oak and have minimum 12 inch shelf width. Bathroom cabinets shall have mirror doors and be set into the wall. There is a $X,XXX allowance for the cabinets.

The cabinets are usually selected after construction starts and the purchaser is given an allowance based on average cabinet cost.

---

(24) Stairs—The stairs and handrail for the stairs shall be oak and be sized to meet the applicable codes. The stairs and railing for the basement steps can be either pine or oak.

The stair material for the basement stairs is usually of a lesser quality than the living area stairs.

---

(25) Finished Floors

Ceramic tile shall be installed at the following locations:

- Foyer
- Kitchen
- Dinette
- Bathrooms

The allowance for ceramic tile is $X,XXX.

Carpet shall be installed at the following locations:

- Bedrooms
- Family room
- Office
- Living room
- Dining room
- Stairs
- Upstairs hallway

The allowance for carpet is $X,XXX.

All flooring is placed on ¾ inch tongue and groove, with ¼ inch underlayment and 4 × 8 sheets of plywood. Plywood is glued to joists.

---

(26) Lighting fixtures—The contractor will install all switches and wiring for lighting fixtures and plug receptacles. The total number of fixtures needed is 20. Fixture allowance is $X,XXX.

(27) Vanities and Countertops—These items will be installed by the contractor with an allowance of $X,XXX.

(28) Appliances—Contractors shall install the following appliances:

- Kitchen range $X,XXX allowance
- Refrigerator $X,XXX allowance
- Dishwasher $X,XXX allowance
- Washer $X,XXX allowance
- Dryer $X,XXX allowance

Items 25 through 28 are typical items with allowances and are self-explanatory.

(29) Air Conditioning Unit—Contractor shall provide a properly sized air conditioner from ABC Air Conditioning.

No explanation needed for air conditioning unit.

SELECTIONS (Form 1)

Project _____

Purchaser _____

Date _____

(1) Family Room
   Floor covering: Type _____ Color _____ Style No. _____

   Fireplace: Face and Hearth: _____

   Material _____ Color _____

   Mantel _____

(2) Kitchen
   Floor Covering: Type _____ Color _____ Style No. _____

   Cabinets: Color _____ Style No. _____

   Countertops: Type _____ Color _____ Style No. _____

   Sink: Type _____ Style No. _____

   Appliances: Refrigerator model _____

   Dishwasher model _____

   Range model _____

   Light fixtures: Type _____ Style No. _____

   Type _____ Style No. _____

(3) Breakfast Room/Dinette
   Floor covering: Type _____ Color _____ Style No. _____

   Light fixtures: Type _____ Style No. _____

(4) Dining Room
   Floor covering: Type _____ Color _____ Style No. _____

   Light fixtures: Type _____ Style No. _____

(5) Living Room
   Floor covering: Type _____ Color _____ Style No. _____

(6) Office/Den
   Floor covering: Type _____ Color _____ Style No. _____

(7) Foyer

       Front door: Type _____   Style No. _____

       Floor covering: Type _____   Color _____   Style No. _____

       Light fixtures: Type _____   Style No. _____

(8) Stairs

       Floor covering: Type _____   Color _____   Style No. _____

(9) Utility Room

       Floor covering: Type _____   Color _____   Style No. _____

       Light fixtures: Type _____   Style No. _____

       Countertops: Type _____   Color _____   Style No. _____

       Commode: Type _____   Color _____   Style No. _____

       Shower: Door _____   Rod _____

                     Wall tile cover _____   Wall tile style no. _____

       Vanity: Color _____   Style No. _____

(10) Master Bath

       Floor covering: Type _____   Color _____   Style No. _____

       Light fixtures: Type _____   Style No. _____

       Countertops: Type _____   Color _____   Style No. _____

       Commode: Type _____   Color _____   Style No. _____

       Shower: Door _____   Rod _____

                     Wall tile color _____   Wall tile style no. _____

       Vanity: Color _____   Style No. _____

(11) Hall Bath

       Floor covering: Type _____   Color _____   Style No. _____

       Light fixtures: Type _____   Style No. _____

       Countertops: Type _____   Color _____   Style No. _____

       Commode: Type _____   Color _____   Style No. _____

       Shower: Door _____   Rod _____

                     Wall tile cover _____   Wall tile style no. _____

       Vanity: Color _____   Style No. _____

(12) Bedrooms

       Floor covering–Master: Type _____   Color _____   Style No. _____

Floor covering—#1:   Type _____   Color _____   Style No. _____

Floor covering—#2:   Type _____   Color _____   Style No. _____

Floor covering—#3:   Type _____   Color _____   Style No. _____

Floor covering—#4:   Type _____   Color _____   Style No. _____

(13) Halls

    Floor covering: Type _____   Color _____   Style No. _____

(14) Utility Room

    Appliances: Washer _____   Style No. _____

                 Dryer_____   Style No. _____

    Floor covering: Type _____   Color _____   Style No. _____

(15) Miscellaneous

    Window trim: Type _____

    Windows: Type _____   Style No. _____

    Facade: Type _____   Color _____   Style No. _____

               Type _____   Color _____   Style No. _____

    Interior doors: Type _____   Style No. _____

    Garage doors: Type _____   Color _____   Style No. _____

    Heat: No. of units _____   Style No. _____

    A/C: No. of units _____   Style No. _____

(16) Exterior of Structure

    Siding: Color _____   Style No. _____

    Trim: Color _____   Style No. _____

    Brick: Color _____   Style No. _____

    Windows: Color _____   Style No. _____

(1) Floor Coverings

Family room: Type _____   Color _____   Style No. _____

Kitchen: Type _____   Color _____   Style No. _____

Breakfast room: Type _____   Color _____   Style No. _____

Dining room: Type _____   Color _____   Style No. _____

Living room: Type _____   Color _____   Style No. _____

Office: Type _____   Color _____   Style No. _____

Foyer: Type _____   Color _____   Style No. _____

Stairs: Type _____   Color _____   Style No. _____

Utility room: Type _____   Color _____   Style No. _____

Master bath: Type _____   Color _____   Style No. _____

Hall bath: Type _____   Color _____   Style No. _____

Bedroom #1:   Type _____   Color _____   Style No. _____

Bedroom #2:   Type _____   Color _____   Style No. _____

Bedroom #3:   Type _____   Color _____   Style No. _____

Bedroom #4:   Type _____   Color _____   Style No. _____

Master bdrm: Type _____   Color _____   Style No. _____

Halls: Type _____   Color _____   Style No. _____

(2) Appliances

Refrigerator: Color _____   Style No. _____

Range: Color _____   Style No. _____

Dishwasher: Color _____   Style No. _____

Dryer: Color _____   Style No. _____

Washer: Color _____   Style No. _____

(3) Counter Tops

Kitchen: Type _____   Color _____   Style No. _____

Master bath: Type _____   Color _____   Style No. _____

Hall bath: Type _____   Color _____   Style No. _____

Powder room: Type _____   Color _____   Style No. _____

(4) Light Fixtures

        Kitchen: Type _____   Style No. _____

        Dining room: Type _____   Style No. _____

        Foyer: Type _____   Style No. _____

        Powder room: Type _____   Style No. _____

        Master bath: Type _____   Style No. _____

        Hall bath: Type _____   Style No. _____

(5) Plumbing Fixtures

        Master bath: Type _____   Style No. _____

        Hall bath: Type _____   Style No. _____

        Powder room: Type _____   Style No. _____

        Kitchen: Type _____   Style No. _____

(6) Cabinets

        Kitchen: Type _____   Color _____   Style No. _____

        Master bath: Type _____   Color _____   Style No. _____

        Hall bath: Type _____   Color _____   Style No. _____

        Powder room: Type _____   Color _____   Style No. _____

(7) Shower Wall Tile

        Utility room: Type _____   Color _____   Style No. _____

        Master bath: Type _____   Color _____   Style No. _____

        Hall bath: Type _____   Color _____   Style No. _____

(8) Fireplace

        Living room: Hearth _____   Mantel _____

        Family room: Hearth _____   Mantel _____

(9) Miscellaneous

        Window trim: Type _____

        Windows: Type _____   Style No. _____

        Facade: Type _____   Color _____   Style No. _____

                Type _____   Color _____   Style No. _____

        Interior doors: Type _____   Style No. _____

        Garage doors: Type _____   Color _____   Style No. _____

        Heat: No. of units _____   Style No. _____

        A/C: No. of units _____   Style No. _____

(10) Exterior of Structure

Siding: Color _____ Style No. _____

Trim: Color _____ Style No. _____

Brick: Color _____ Style No. _____

Windows: Color _____ Style No. _____

Appendix A

Contract Sample

This appendix contains a sample construction contract. This contract is only a sample and is not meant to be used unless an attorney is consulted to explain and make the necessary modifications. The authors and publisher are not responsible for the unauthorized use of this sample contract. Any party using this appendix agrees to hold the authors and publisher harmless for any and all causes of action.

Residential Construction Contract

Purchaser: _____

Address: _____

_____

_____

Phone: _____

General Contractor: _____

Address: _____

_____

_____

Phone: _____

The general contractor agrees to construct a _____

style home with basement and _____ garage, subject to the terms and conditions of this

contract, at

_____ , whose legal description is: __

_____

_____

_____

_____

The purchaser agrees to purchase the residence at this address under the terms and conditions of this contract.

Contract Documents—The following items are hereby included as part of this contract:

- Design drawings: Title:

    Prepared by:

- Attached Specifications

- Local Building Codes

- Occupational Health and Safety Act (OSHA) Requirements

- Local Fire Codes

- All Applicable Material Codes

In the event contract documents conflict, the order of precedence shall be as listed above.

Property Ownership—The purchaser represents that the property is solely and exclusively theirs. The purchaser further represents that the property is free and clear of all liens and encumbrances unless noted elsewhere in this contract.

Taxes and Assessments—The purchaser and general contractor shall each pay their pro-rata share of real estate taxes and assessments for the year of closing at the time of closing. The estimated real estate taxes for the year of proration shall be based on 105% of the previous year. The general contractor shall pay any taxes accrued prior to the year of closing.

Purchase Price and Payouts:

Base Amount $

Extras $

Total $

Schedule of Payments:

 (i) At contract signing—

(ii) House under roof—

(iii) Interior ready for paint—

(iv) Completion—

Loan Commitment—The purchaser must receive a loan commitment within 30 days from signing of this contract in the amount of $_____ . The purchaser shall make every reasonable effort to obtain the loan commitment and shall pay all usual and customary charges imposed by the lending institution. The purchaser shall provide a copy of the commitment letter to the general contractor within the 30 days. If the purchaser is unable to obtain financing, the purchaser shall notify the general contractor within the 30 days. If the purchaser notifies the general contractor that a loan commitment cannot be obtained, the general contractor has the option to obtain a loan commitment on behalf of the purchaser. All associated fees will be paid for by the purchaser. Failure of the general contractor to exercise this option makes this contract null and void.

Permits—The purchaser shall pay all fees for any and all permits required for the project, including the building permit. In the event that the design drawings are not acceptable for the obtaining of a building permit, it shall be the purchaser's responsibility to have the drawings modified as necessary at no cost to the general contractor. Other permits include but are not limited to building, water, sewer, septic, well, and site grading. The general contractor shall be responsible for obtaining the permits.

Testing—The purchaser is responsible for obtaining, as well as paying all fees for any and all testing needed for construction of the residence. Testing includes, but is not limited to, soil strength analysis, permeability tests, and water quality testing. The general contractor shall arrange for all testing with the exception of the soil test, which shall be arranged by the owner with an analysis received prior to the start of construction.

Time of Construction—The structure shall be completed no later than six months from the signing of the contract. The six month period shall be extended for circumstances beyond the control of the general contractor such as, but not limited to labor strikes, material availability, war, natural disasters, and other unforeseen acts of God. The six month period shall be extended by the length of time of the interruption. The general contractor agrees to pay $_____ per day for every day beyond the six months that the structure is not completed. This $_____ is compensation for inconveniences and living costs for the purchaser. The purchaser agrees to close on the house, pay the full purchase price, and receive ownership no later than 14 days after completion.

Quality of Work—All work performed by the general contractor and subcontractors hired by the general contractor shall be in accordance with good construction practice, drawings, and specifications. The work shall meet all applicable building codes and be performed in a quality manner. All work shall be free of defects and faults.

Materials—The contractor is required to furnish all material necessary to complete the house. The contractor is the owner of the material, and is therefore responsible for the material until it is installed. The material furnished shall be of good quality, new and at least equal to the quality of the standard in the industry for the respective type of product. The general contractor shall be allowed to substitute material when the specified material is no longer available or is not feasible to purchase. The purchaser shall be notified of such changes and be allowed a reasonable objection to the substitute.

Change Orders—Any changes, deviation, or additional work shall only be performed after a change order has been received by the general contractor. The change order shall contain a description of the work, change in price, and the signature of the purchaser. The change order shall be signed by the general contractor if the general contractor agrees to perform the work at the stipulated price. The general contractor shall not have the right to refuse any reasonable change order. In the case where it is not possible to wait for a properly processed change order, the general contractor shall notify the purchaser of the additional work and price for the change in work. The owner shall provide a verbal response and then provide a properly processed change order within four working days.

Purchaser Insurance—The purchaser shall obtain and pay for fire insurance with extended coverage and builders risk insurance for joint coverage of the purchaser, general contractor, and subcontractors. The insurance shall be in the amount of the contract price of the structure excluding concrete work items. The purchaser shall furnish the general contractor evidence of this insurance which shall be obtained under the condition that the insurance cannot be cancelled without 30 days notice to the general contractor.

General Contractor—The general contractor shall obtain and pay for Comprehensive General Liability Insurance and all other insurance required by law. These insurance requirements include but are not limited to:

- Workman's Compensation
- General Liability—Personal Injury
- General Liability—Property Damage

The general contractor shall provide statutory amounts for insurance required by law and $_____ each person and $ _____ each occurrence for general liability insurance. Prior to the start of construction, the general contractor shall provide evidence of the required insurance with a provision that the policy cannot be cancelled without 30 days notice to the purchaser. The policy must cover all material prior to attachment to the structure . All subcontractors employed by the general contractor must maintain insurance in the same amount as required of the general contractor.

Warranty provided by General Contractor

Alternative 1—The general contractor warrants all work against defects. The general contractor agrees to correct any defective work at no cost to the purchaser. This warranty is in effect only for one year from the date of completion or date of possession, whichever comes first.

Warranties Provided by Statute—The purchaser is limited to warranties provided by the general contractor as discussed above. The general contractor makes no other warranties, express or implied, including, but not limited to, the implied warranty of habitability or merchantability or fitness for a particular purpose.

Safety—The general contractor shall be solely responsible for providing a safe work site. The work site shall be safe for workers and all persons. The general contractor shall be responsible for providing a safe work site and initiating safety programs for all subcontractors. Precautions shall be taken to protect all property. The general contractor is responsible for meeting all federal and state safety requirements including compliance with OSHA regulations.

Subcontractors—The general contractor is responsible for insuring that all work performed by subcontractors is performed in an acceptable manner and in accordance with the contract, design drawings, and specifications. The general contractor shall not use any subcontractors who are not fully insured as required by this contract or any subcontractor reasonably objected to by the purchaser. The contractor shall not be required to hire any specific subcontractor unless expressly specified in this contract.

Destruction of Premises—In the event the structure is destroyed before the purchaser has taken ownership, the general contractor shall choose one of the following options.

Within two weeks of the date of occurrence of the destruction, the general contractor shall:

(i) Rebuild and finish the structure within 180 calendar days using the insurance proceeds, or

(ii) Void the contract, receive the insurance proceeds, and return deposit to purchaser.

Utilities—The general contractor shall be responsible for arranging for permanent utility connections to the residence as well as for the cost incurred to provide utility service to the residence. All costs incurred for utility services prior to the closing shall be paid for by the general contractor. The general contractor shall notify all utility companies prior to any digging to avoid damage to buried service lines.

Clean Up—The general contractor shall keep the construction site in a neat and clean condition during the life of this contract regardless if the work is being performed by the general contractor or a subcontractor. All material shall be stored in an orderly, neat, and safe manner. The entire premises, interior and exterior, shall be cleaned immediately prior to closing, including removal of all trash and debris.

Punch List—At the discretion of the general contractor, the purchaser may take possession of the property upon substantial completion of the residence. If the purchaser does take possession a final "punch list" shall be recorded. The punch list shall list with specificity, what work is needed to achieve final completion of the contract. The contractor agrees to work diligently to complete the punch list items within 30 days of possession. However, the general contractor shall be paid in full regardless of the unfinished items on the punch list.

Assignment of Contract—This contract cannot be assigned to any other party without prior approval from the nonassigning party. However, this contract shall be binding upon and be for the benefit of the parties to the agreement, their heirs and successors and personal representatives.

Default—In the event that the contractor fails to carry out the terms of this contract the purchaser shall make a written notice to the contractor of the lack of progress. If the contractor does not respond in 10 working days, the purchaser may terminate the contract. The general contractor shall pay to the purchaser any additional cost beyond the original contract for completion of the structure. In the event the purchaser fails to make a required payment the contractor shall make a written protest of the missed payment. If the purchaser does not make payment, the contractor may void the contract and receive payment for all work completed. The nonbreaching party shall be compensated for any losses including attorney's fees for the breach.

Entire Agreement—This agreement constitutes the entire agreement and is a final complete expression of the agreement between the parties. All prior discussions, promises, or representations are merged into this document.

Appendix B

Contract Specifications

This appendix contains a sample set of specifications. This appendix is only a sample and is not meant to be used without modification. Any party using this appendix assumes all the risks and agrees to hold the authors and publisher harmless for any and all causes of action.

Excavation—The contractor shall perform the following excavation at the contractors cost:

• Removal of tree or bush growth in area of excavation.

• Strip topsoil and stockpile.

• All necessary foundation excavating.

• All necessary foundation backfilling.

• Spreading of stockpiled soil upon completion of the structure.

The purchaser shall pay the expense for any labor or material for the following items:

• Removal of trees or bush growth outside excavation area

• Additional grading beyond spreading of topsoil.

• Removal and disposal of any buried items.

Concrete Foundations—All footings and foundations shall reach at least _____ psi in 28 days. Steel reinforcing bars shall be _____ in. diameter and installed at locations shown on design drawings.

Steel Members—Steel pipe columns shall be _____ in. outside diameter and filled with concrete. The steel pipe columns shall conform to ASTM standards.

Steel beams shall be _____ or as specified on the design drawings. Steel beams shall conform to ASTM standards.

Foundation Wall Damproofing—All portions of the foundation wall that will be buried below grade shall have a coating of an asphalt based foundation damproofing on the exterior face. Damproofing shall be applied when the temperature is above _____degrees F and no rain or snow will occur within _____ hours. Damproofing shall consist of two layers applied in opposite directions.

Footing Drains—Footing drains shall be a minimum of _____ in. diameter plastic perforated pipe. The pipe shall meet ASTM standards. The drain shall be placed in _____ in. of stone. The footing drain shall empty by gravity away from the structure or into a sump pit.

Fireplace—Fireplaces shall be prefabricated with the size shown on the design drawings. Fireplace design shall be certified by the American Gas Association and have appropriate venting.

Exterior Walls—Exterior walls shall be constructed of _____ studs unless noted differently on the drawing. Exterior cladding shall be cedar siding or brick over wood sheathing as shown on the design drawings. A building wrap shall be used between the cladding and the sheathing. Cedar siding shall be stained with two coats.

Interior Walls—Interior walls shall be constructed of _____ studs that are ___ in. on center.

Floor and Ceiling Framing—All wood framing shall be 2 X 8, 2 X 10, 2 X 12 and spaced as noted on the drawings. Wood shall be _____. Joists shall have bracing located between adjacent joists.

Prefabricated Wood Trusses—Prefabricated trusses shall be designed to carry the appropriate loads. A stamp by a licensed structural engineer shall accompany each truss.

Roofing—Shingles shall be _____lbs. fiberglass with a one year warranty on labor and a minimum 25 year warranty on the material. Felt under shingles shall be a minimum of _____ pounds. The roofing material shall be secured to _____in. thick exterior grade plywood.

Gutters and Downspouts—All gutters and downspouts shall be __ in. aluminum box type construction.

Windows—All windows shall be casement with aluminum-clad exterior and shall be manufactured by _____ windows or an approved equivalent. Window shall include screen and storm windows.

Garage Doors—All garage doors shall be plain raised redwood panel doors. Redwood doors shall be weather treated.

Septic System—Complete septic shall be installed as per plans, including stone, filter paper, pipe lines, septic tanks and backfilling. All work shall meet county approval.

Driveway—The driveway shall consist of a _____ in. stone base below ___ in. thick asphalt. The stone shall be compacted prior to placing the asphalt.

Concrete Slabs—All concrete slabs shall be poured on a minimum of __ in. of gravel and visqueen. The concrete slabs shall be reinforced with welded wire fabric. The concrete mix shall be such that the strength of the concrete is at least _____ psi within 28 days.

Plumbing—The general contractor shall provide the following:

Water:
- Piping for all sink, bathrooms, dishwasher, water heaters, and all other items that require water.
- ____ gallon hot water heater
- Well
- Kitchen sink
- Utility sink
- Shower bases
- Faucets
- Exterior faucet
- Sump pumps
- Water softener

Wastewater:
- Piping from all sinks, bathroom, etc.
- Septic system

Gas:
- Gas piping from exterior source to furnaces, stoves, fireplace log lighter, dryer, and water heater.

Allowances:
Septic $

Water softener $

Well $

Plumbing fixtures $

Heating—Two high efficiency gas burning furnaces with galvanized metal supply ducts shall be sized and installed. The furnace shall be _____ brand or at least of equal quality. Furnace shall be operable by digital thermometer with programming capability.

Insulation—Insulation with the following R number shall be provided.

Ceiling   R =

Walls   R =

Floor   R =

Insulation shall only be provided in garage walls common to living areas. The exterior of the house shall be wrapped with draft reducing plastic.

Paint—Two coats of flat latex paint shall be applied to all walls and ceiling. Colors other than white shall result in an extra charge.

Trim—All interior doors and trim shall be _____. Doors shall be stained and sealed and of the six-panel style. Base trim and door casings as well as any crown moldings specified on drawings shall be provided by the contractor and shall be stained and sealed.

Cabinets—Kitchen cabinets shall be constructed of oak and have minimum 12 in. shelf width. Bathroom cabinets shall have mirror doors and be set into the wall. There is a $_____ allowance for the cabinets.

Stairs—The stairs and handrail for the stairs shall be oak and be sized to meet the applicable codes. The stairs and railing for the basement steps can be either pine or oak.

Finished Floors

Ceramic tile shall be installed at the following locations:

- Foyer

- Kitchen

- Dinette

- Bathrooms

The allowance for ceramic tile is $_____.

Carpet shall be installed at the following locations:

- Bedrooms

- Family room

- Office

- Living room

- Dining room

- Stairs

- Upstairs hallway

The allowance for carpet is $_____.

Lighting fixtures—The contractor will install all switches and wiring for lighting fixtures and plug receptacles. The total number of fixtures needed is ___. Fixture allowance is $_____.

Vanities and Countertops—These items will be installed by the contractor with an allowance of $_____.

Appliances—Contractors shall install the following appliances:

- Kitchen range $_____ allowance

- Refrigerator $_____ allowance

- Dishwasher $_____ allowance

- Washer $_____ allowance

- Dryer $_____ allowance

Air Conditioning Unit—Contractor shall provide a properly sized air conditioner from

_____.

Project _____

Purchaser _____

Date _____

(1) Family Room
     Floor covering: Type _____   Color _____   Style No. _____

     Fireplace: Face and Hearth: _____

                Material ____   Color _____

                Mantel _____

(2) Kitchen
     Floor Covering: Type _____   Color _____   Style No. _____

     Cabinets: Color_____   Style No. _____

     Countertops: Type _____   Color _____   Style No. _____

     Sink: Type _____   Style No. _____

     Appliances: Refrigerator model _____

                Dishwasher model _____

                Range model _____

     Light fixtures: Type _____   Style No. _____

                Type _____   Style No. _____

(3) Breakfast Room/Dinette
     Floor covering: Type _____   Color _____   Style No. _____

     Light fixtures: Type _____   Style No. _____

(4) Dining Room
     Floor covering: Type _____   Color _____   Style No. _____

     Light fixtures: Type _____   Style No. _____

(5) Living Room
     Floor covering: Type _____   Color _____   Style No. _____

(6) Office/Den
     Floor covering: Type _____   Color _____   Style No. _____

(7) Foyer

    Front door: Type _____   Style No. _____

    Floor covering: Type _____   Color _____   Style No. _____

    Light fixtures: Type _____   Style No. _____

(8) Stairs

    Floor covering: Type _____   Color _____   Style No. _____

(9) Utility Room

    Floor covering: Type _____   Color _____   Style No. _____

    Light fixtures: Type _____   Style No. _____

    Countertops: Type _____   Color _____   Style No. _____

    Commode: Type _____   Color _____   Style No. _____

    Shower: Door _____   Rod _____

    Wall tile cover _____   Wall tile style no. _____

    Vanity: Color _____   Style No. _____

(10) Master Bath

    Floor covering: Type _____   Color _____   Style No. _____

    Light fixtures: Type _____   Style No. _____

    Countertops: Type _____   Color _____   Style No. _____

    Commode: Type _____   Color _____   Style No. _____

    Shower: Door _____   Rod _____

    Wall tile color _____   Wall tile style no. _____

    Vanity: Color _____   Style No. _____

(11) Hall Bath

    Floor covering: Type _____   Color _____   Style No. _____

    Light fixtures: Type _____   Style No. _____

    Countertops: Type _____   Color _____   Style No. _____

    Commode: Type _____   Color _____   Style No. _____

    Shower: Door _____   Rod _____

    Wall tile cover _____   Wall tile style no. _____

    Vanity: Color _____   Style No. _____

(12) Bedrooms

    Floor covering–Master: Type _____   Color _____   Style No. _____

Floor covering—#1:   Type _____   Color _____   Style No. _____

Floor covering—#2:   Type _____   Color _____   Style No. _____

Floor covering—#3:   Type _____   Color _____   Style No. _____

Floor covering—#4:   Type _____   Color _____   Style No. _____

(13) Halls

    Floor covering: Type _____   Color _____   Style No. _____

(14) Utility Room

    Appliances: Washer _____   Style No. _____

                Dryer _____   Style No. _____

    Floor covering: Type _____   Color _____   Style No. _____

(15) Miscellaneous

    Window trim: Type _____

    Windows: Type _____   Style No. _____

    Facade: Type _____   Color _____   Style No. _____

                 Type _____   Color _____   Style No. _____

    Interior doors: Type _____   Style No. _____

    Garage doors: Type _____   Color _____   Style No. _____

    Heat: No. of units _____   Style No. _____

    A/C: No. of units _____   Style No. _____

(16) Exterior of Structure

    Siding: Color _____   Style No. _____

    Trim: Color _____   Style No. _____

    Brick: Color _____   Style No. _____

    Windows: Color _____   Style No. _____

(1) Floor Coverings

Family room: Type _____ Color _____ Style No. _____

Kitchen: Type _____ Color _____ Style No. _____

Breakfast room: Type _____ Color _____ Style No. _____

Dining room: Type _____ Color _____ Style No. _____

Living room: Type _____ Color _____ Style No. _____

Office: Type _____ Color _____ Style No. _____

Foyer: Type _____ Color _____ Style No. _____

Stairs: Type _____ Color _____ Style No. _____

Utility room: Type _____ Color _____ Style No. _____

Master bath: Type _____ Color _____ Style No. _____

Hall bath: Type _____ Color _____ Style No. _____

Bedroom #1: Type _____ Color _____ Style No. _____

Bedroom #2: Type _____ Color _____ Style No. _____

Bedroom #3: Type _____ Color _____ Style No. _____

Bedroom #4: Type _____ Color _____ Style No. _____

Master bdrm: Type _____ Color _____ Style No. _____

Halls: Type _____ Color _____ Style No. _____

(2) Appliances

Refrigerator: Color _____ Style No. _____

Range: Color _____ Style No. _____

Dishwasher: Color _____ Style No. _____

Dryer: Color _____ Style No. _____

Washer: Color _____ Style No. _____

(3) Counter Tops

Kitchen: Type _____ Color _____ Style No. _____

Master bath: Type _____ Color _____ Style No. _____

Hall bath: Type _____ Color _____ Style No. _____

Powder room: Type _____ Color _____ Style No. _____

(4) Light Fixtures

Kitchen: Type _____   Style No. _____

Dining room: Type _____   Style No. _____

Foyer: Type _____   Style No. _____

Powder room: Type _____   Style No. _____

Master bath: Type _____   Style No. _____

Hall bath: Type _____   Style No. _____

(5) Plumbing Fixtures

Master bath: Type _____   Style No. _____

Hall bath: Type _____   Style No. _____

Powder room: Type _____   Style No. _____

Kitchen: Type _____   Style No. _____

(6) Cabinets

Kitchen: Type _____   Color _____   Style No. _____

Master bath: Type _____   Color _____   Style No. _____

Hall bath: Type _____   Color _____   Style No. _____

Powder room: Type _____   Color _____   Style No. _____

(7) Shower Wall Tile

Utility room: Type _____   Color _____   Style No. _____

Master bath: Type _____   Color _____   Style No. _____

Hall bath: Type _____   Color _____   Style No. _____

(8) Fireplace

Living room: Hearth _____   Mantel _____

Family room: Hearth _____   Mantel _____

(9) Miscellaneous

Window trim: Type _____

Windows: Type _____   Style No. _____

Facade: Type _____   Color _____   Style No. _____

                   Type _____   Color _____   Style No. _____

Interior doors: Type _____   Style No. _____

Garage doors: Type _____   Color _____   Style No. _____

Heat: No. of units _____   Style No. _____

A/C: No. of units _____   Style No. _____

(10) Exterior of Structure

Siding: Color _____ Style No. _____

Trim: Color _____ Style No. _____

Brick: Color _____ Style No. _____

Windows: Color _____ Style No. _____

Index

Illustration page numbers are in **boldface.**

DISK WARRANTY